W9-DJD-829

Engineering Materials and Processes

Series Editor

Professor Brian Derby, Professor of Materials Science
Manchester Materials Science Centre, Grosvenor Street, Manchester, M1 7HS, UK

Other titles published in this series

Fusion Bonding of Polymer Composites
C. Ageorges and L. Ye

Composite Materials
D.D.L. Chung

Titanium
G. Lütjering and J.C. Williams

Corrosion of Metals
H. Kaesche

Corrosion and Protection
E. Bardal

Intelligent Macromolecules for Smart Devices
L. Dai

Microstructure of Steels and Cast Irons
M. Durand-Charre

Phase Diagrams and Heterogeneous Equilibria
B. Predel, M. Hoch and M. Pool

Computational Mechanics of Composite Materials
M. Kamiński

Gallium Nitride Processing for Electronics, Sensors and Spintronics
S.J. Pearton, C.R. Abernathy and F. Ren

Materials for Information Technology
E. Zschech, C. Whelan and T. Mikolajick

Fuel Cell Technology
N. Sammes

Casting: An Analytical Approach
A. Reikher and M.R. Barkhudarov

Computational Quantum Mechanics for Materials Engineers
L. Vitos

Modelling of Powder Die Compaction
P.R. Brewin, O. Coube, P. Doremus and J.H. Tweed

Silver Metallization
D. Adams, T.L. Alford and J.W. Mayer

Reza Javaherdashti

Microbiologically Influenced Corrosion

An Engineering Insight

Reza Javaherdashti, PhD
Extrin Consultants
6 McElligott Court
Canning Vale
Western Australia 6155
Australia

ISBN 978-1-84800-073-5 e-ISBN 978-1-84800-074-2

DOI 10.1007/978-1-84800-074-2

Engineering Materials and Processes ISSN 1619-0181

British Library Cataloguing in Publication Data
Javaherdashti, Reza
Microbiologically influenced corrosion : an engineering insight. - (Engineering materials and processes)
1. Microbiologically influenced corrosion
I. Title
620.1'1223
ISBN-13: 9781848000735

Library of Congress Control Number: 2008921915

Cover design: eStudio Calamar S.L., Girona, Spain

Printed on acid-free paper

9 8 7 6 5 4 3 2 1

springer.com

Preface

A Few Words About the Structure of This Book

Let me be honest with you: I rarely read the prefaces of books! It is, I guess, because I think that I am interested only in the context of the book, not what the author wants to teach me about how to read the book. I have found very few books whose "introduction" has been interesting to me. But for this book, I strongly recommend that readers study this preface to understand why I chose the structure used in placing the chapters.

I have always wanted to write a book about microbial corrosion (there are some alternative names to address this type of corrosion; they're given in Chapter 4) that would have a rather wide audience, ranging from academics (lecturers, researchers, postgraduate students) to industry specialists (field engineers, design engineers, industry managers). This goal may seem very enthusiastic, to put it politely. There has always been an unseen, undefined gap between different disciplines of science (Videla has touched on this very meaningfully, and his paper is quoted in later chapters of this book), let alone between industry and research/university environments.

Despite these obstacles, I have tried to be fair to both of my target audiences, university/research and industry. In addition, I have tried to focus on a very important aspect of corrosion mitigation: the human management factor. With this in mind, readers must understand the structure of this book to gain the maximum advantage of reading a book with such a wide potential audience.

We will start our microbial corrosion journey by reviewing some basic corrosion – to put it more precisely, electrochemical corrosion, in Chapter 1. In this chapter, some basic facts regarding electrochemical corrosion are reviewed to a limited extent that may be useful to understand the logic behind using methods and techniques such as cathodic protection, coating and use of inhibitors which are explained in Chapter 2 in the section "Technical Mitigation of Corrosion". This much can be found in almost every book written on corrosion or microbial corrosion, where basic information regarding corrosion and its mitigation is given. However, a very important part of mitigation methods against corrosion (and, therefore, against microbial corrosion) is the factor of human management; no matter how good the techniques are that we use in combating microbial corrosion, if there is poor communication between the technical staff (engineers, technicians, foremen, *etc.*) and the management, the resultant practice will have very limited impact on upgrading the performance of the system. If management cannot understand the importance of microbial corrosion, even the best corrosion engineer cannot justify the expense of microbial corrosion recognition and treatment. This may not be a serious matter for academic researchers, but it certainly is important for both industry researchers and field engineers. Chapter 3 deals with a very genuine and innovative concept called "corrosion knowledge management (CKM)" to differentiate it from what is normally known as "corrosion management". While the later refers to the technicalities involved in corrosion treatment (such as the best design and practice of cathodic protection, the choice of inhibitors and coatings and the like), corrosion knowledge management concentrates more on managerial aspects. Therefore, although a manager may not know what a reference electrode is for, or what the difference between an inhibitor and a non-oxidising biocide is, this manager will need to know how, economically and environmentally, microbial corrosion in particular and corrosion in general could be dangerous. A manager also needs to have a managerial system in place so that an organisational chart can be defined. Chapter 3 introduces the basics of such managerial needs.

Chapter 4 may be described as the heart of this book. It begins with a historical profile of microbial corrosion and definitions, followed by topics such as the "paradoxical" effect of biofilms on corrosion. The text continues with a review of some types of bacteria which are of interest in the microbial corrosion literature. Some of these bacteria, such as the sulphate-reducing bacteria (SRB) have been long known to researchers and industry. Some, like the iron-reducing bacteria (IRB) are not that well known, and I have dubbed them as "shy" as they seem not to get the attention of researchers the way SRB do. The possible role of magnetic bacteria in corrosion is stated for the first time in the literature of microbial corrosion, to the best of my knowledge. Magnetic bacteria are very interesting, and they form an "exotic" realm for further research. Chapter 4 also includes some important concepts regarding the possible impacts and effects of SRB and IRB on enhancing stress corrosion cracking.

Before closing this summary of Chapter 4, I want to add a few words about SRB. I do agree with Brenda Little and Patricia Wagner in calling the importance of SRB a "myth" of microbial corrosion research and practice. But, readers may wonder, if the importance of these bacteria has been naively exaggerated, why I am allocating so many pages to explain them? The answer is easy: the stronger a wrong belief is, the more you have to explain it to make it clear. I have tried to explain that although SRB are important, they are not so important as to cause us to forget other types of bacteria involved in microbial corrosion.

Chapter 5 considers what and how factors must come together to put a system in danger of microbial corrosion. This chapter studies the effect of water quality and velocity, oxygen, hydrotesting and other relevant factors in the initiation of microbial corrosion in *any* system that has the potential. It does not matter which industry the system may belong to. As long as the required factors are in place, the system will become vulnerable.

Chapter 6 studies the parameters required for "recognition" of microbial corrosion, factors such as the shape of the pits, mineralogical "fingerprints", and the appearance of corrosion products. This chapter ends with a review of "detection" techniques which are basically microbiological and electrochemical. Thus, for example, culturing, molecular biological methods, and rapid check tests and their pros and cons are among the topics that are. An important part of this chapter for researchers is the review of electrochemical methods and their importance in microbial corrosion investigations.

In Chapter 7, I try to show that microbial corrosion can have more or less similar patterns despite different systems in which it is occurring. This chapter shows how microbial corrosion in fire water lines could be similar to that happening within the legs of a submersible off-shore platform, and how buried pipelines and steel piles of a jetty could experience almost the same scenarios of microbial attack.

Almost no engineering material is safe from or immune to microbial corrosion. In Chapter 8 the vulnerability and susceptibility of copper and cupronickels, duplex stainless steels and concrete will be discussed in a brief and informative manner. I had my reasons for picking these materials: copper and its alloys have the reputation of being poisonous to micro-organisms, duplex stainless steels are known for their high resistance to corrosion thanks to their duplex microstructures of ferrite and austenite, and concrete is widely used in both the marine and water industries because of its good performance and cost effectiveness.

Having said so much about microbial corrosion, in Chapter 9 I address a logical expectation: how is this type of corrosion treated? I go through only the physical-mechanical, chemical, biological, and electrochemical (including cathodic protection) means and factors that have been used thus far to treat and mitigate microbial corrosion. An interesting point, among others, could be the possible explanation of why cathodic protection could be effective (or sometimes ineffective) on microbial corrosion. Although principles of CKM are also applicable here, for reasons that I discussed briefly in the footnote of the opening page of Chapter 9, I did not include the principles in the contents of the chapter.

I have been careful to use language which is very precise, technically sound, and accurate, yet somewhat casual and not too technical. I believe that if there is a truth, it can be explained with accurate yet simple words.

These have been my aims and dreams, and I do hope that my readers will share them with me!

Perth, Australia, 2007 *Reza Javaherdashti*

Acknowledgements

I thank the many people who have helped me through my years spent on research, consulting, and teaching microbial corrosion. My dear friends Professor Hector A. Videla (University of La Plata, Argentina), Professor Dr. Filiz Sarioglu (Middle East Technical University, Turkey), Dr. M. Setareh (Arak University of Medical Sciences, Iran), Mr. R.A. Taheri (University of Tehran, Iran), and Dr. M.M. Vargas (Monash University, Australia) are thanked for their sincere friendship, kindness, and scientific generosity. The support from Dr. Elena Pereloma, Dr. Raman Singh, and Professor Brian Cherry (Monash University, Clayton Campus) is highly appreciated regarding my research on microbial corrosion at Monash University. In this regard, I should also highly thank my good friend Mr. Chris Panter (Monash University, Gippsland Campus) for his priceless assistance, support and contributions. I deeply appreciate my friends and colleagues, Dr. M.S. Parvizi (Principal Technology Engineer, Foster Wheeler Energy Ltd., UK), the recipient of the NACE 2007 Technical Achievement Award, Dr. B. Kermani (KeyTech, UK) and Dr. A. Morshed (Principal Corrosion Engineer, Production Services Network, UK) for reading some parts of this book and providing their useful comments. I am indeed very thankful to a great friend and mentor, Dr. Peter Farinha, Director and Principal Engineer of Extrin Corrosion Consultants (Perth-Australia). I am grateful to him because he believed in me and gave me the opportunity to develop my vision. He was very kind to proofread the manuscript and give suggestions for its improvement. Mrs. Maria Farinha (Director, Extrin) also kindly proofread some of the chapters of this book as a manuscript. There are so many other professionals and friends to whom I am sincerely thankful, but there is not enough space here to express my great feelings and thanks for them.

My parents and my sister are the ones who believed in me not just as a son and a brother but as someone who can make a change. Margie and David Mills are among the beloved ones to whom I want to present my sincere thanks and love. However, the person who agreed to accompany me through all the ups and downs of my life, my soul-mate Asal, has to be the very person to whom I should say a big "thank you". I also hope that my little daughter, Helya, will find this book useful too, someday in the future, perhaps!

There is no doubt that there are many nameless "soldiers" who, both as "white collars" and "blue collars" are fighting globally against corrosion to make life easier for all of us. I would like to dedicate this book to all the scientists and engineers, researchers and technicians, academics and students, managers and administrators who, directly or indirectly, work to fight against a silent serial killer called *Corrosion*!

Perth, Australia, July 2007 *Reza Javaherdashti*

Contents

Chapter 1
A Short Journey to the Realm of Corrosion

1.1 Introduction

This chapter deals with the principles and basics of corrosion to an extent that is necessary for understanding microbial corrosion, or whatever you call it![1]

1.1.1 Definition of Corrosion

Corrosion according to ISO 8044 standard, is defined as "Physicochemical interaction (usually of an electrochemical nature) between a metal and its environment which results in changes in the properties of the metal and which may often lead to impairment of the function of the metal, the environment, or the technical system of which these form a part" [1]. In a sense, corrosion can be viewed as "the chemical reversion of a refined metal to its most stable energy state" [2]. During extractive processes to obtain metals out of their ores or mineral compounds, reductive processes are applied. In these processes, by giving more electrons to metallic compounds in the ore, thermodynamically stable metal in the ore is brought into a thermodynamically instable state by reducing processes of extractive metallurgy. In other words, by investing energy to convert the ore to metal, chemical bonds are broken; oxygen, water, and other anions are removed and the pure metal is arranged in an ordered lattice whose formation requires certain amount of excess energy, different for each metal, to be stored. It is the dissipation of this stored energy that drives the corrosion reaction. As a result, metals always are expected to reach a stable energy level by giving off additional electrons they have received during extractive metallurgical processes. This builds up the thermodynamic basis of oxidation, or more generally termed, corrosion, in metals.

---

[1] See Chapter 4 for alternative names.

The main components of electrochemical corrosion are the Anode (where anodic reactions occur), the Cathode (where cathodic reactions of receiving electrons from the anodic reactions happen), and the aqueous solution, or Electrolyte, which contains positively and negatively charged ions and is a conductor.

As all of the corrosive processes related to micro-organisms occur electrochemically, *i.e.*, in aqueous environments, this chapter will focus on the mechanisms of this type of corrosion and will not explain other types of corrosion such as high-temperature corrosion.

1.2 Electrochemical Corrosion: A Brief Introduction

Definitions of the anode and the cathode are among basic definitions in electrochemical corrosion. The area of the metal surface that corrodes (*i.e.*, where the metal dissolves and goes into solution) is called the anode. The cathode is the area of the metal surface that does not dissolve. In the literature of electrochemistry, reduction and oxidation reactions are defined as when metals lose electrons (*i.e.*, oxidation) or gain electrons (reduction):

$Fe \rightarrow Fe^{2+} + 2e^{-}$

Iron atom → Iron ion (ferrous) + electrons (1.1)

Reaction (1.1) is an example of an oxidation reaction, the oxidation of iron. As it is seen, this reaction gives off electron into the solution. Such reactions are also called anodic reactions.

$2H^{+} + 2e^{-} \rightarrow H_2\uparrow$

Hydrogen ions + electrons → hydrogen gas (1.2)

Reaction (1.2) is an example of a reduction reaction. Reduction reactions are also called cathodic reactions. So, alternatively, one can define corrosion as anodic reactions that are occurring at the anode. Thus, three main components of any electrochemical corrosive reaction are the anode, the cathode, and the solution in which corrosion occurs. This solution is called the electrolyte, and water is always an integral part of it. One may show these components as three corners of an "electrochemical triangle" (see Figure 1.1).

The "electrochemical triangle" implies that for *corrosion to happen, all three components must be available and interactive.* It follows that any method to be implemented to solve a corrosion problem must try to remove at least one of the sides of the triangle. This point will be discussed in more detail later.

Some important points about corrosion are:

- As is evident from definitions of anodic and cathodic reactions, there is exchange of electrons either as liberated (anodic reaction) or gained (cathodic re-

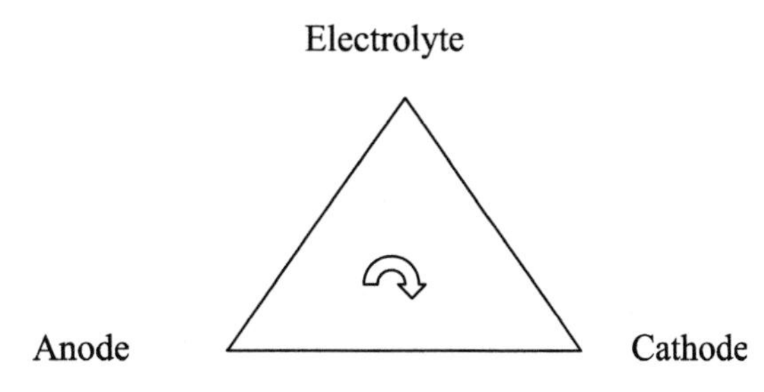

Figure 1.1 Main components of an electrochemical triangle

action). This means that one can actually talk about a flow of electrons, or, the current. On the other hand, if a certain area "A" of the anode loses (or, as cathode, gains) "n" electrons, it follows that an area of "2A" will give (or, again as cathode, will gain) "2n" electrons. So, it is possible to talk about current density instead of current. Defining the current density as total current (in Ampere) passing through an electrolyte per unit area of an electrode, it is shown as A/m^2. A very important point is that, by applying Faraday's laws and noticing that as for many metals of engineering interest the ratio of the relative molar mass (equivalent weight) to the density is roughly constant, then:

$$A/m^2 = mm/yr$$

- The relation states that the corrosion rate of metal, expressed in millimetre per year (mm/yr) is numerically equal to the current density expressed in Ampere per unit area (A/m^2) [3].
- Most of the time, there is only one anodic reaction. For instance, in the "Galvanic cells" where there are two dissimilar metals such as iron and copper, the anodic reaction is always dissolution of the metal with a higher tendency to corrode. However, there could be more than one cathodic reaction. Some of important cathodic reactions are as follows:

  – Reduction of oxygen in neutral/alkaline solutions:

$$O_2 + 2H_2O + 4e^- \rightarrow 4OH^- \tag{1.3}$$

  – Reduction of hydrogen that in anaerobic (oxygen-free) or acid electrolytes switches to:

$$H^+ + e^- \rightarrow H \tag{1.4}$$

Or

$$H_3O^+ + e^- \rightarrow \frac{1}{2}H_2 + H_2O$$

$$H + H \rightarrow H_2 \tag{1.5}$$

This much slower reaction (except in acid conditions) explains why corrosion is hardly seen in anaerobic environments (except with passive/active metals such as stainless steels and in the presence of the micro-organisms) [2].

1.3 When Does Corrosion Happen?

It is of interest to anticipate under what conditions corrosion can actually occur. Every electrochemical reaction has a specific voltage that can be used with respect to a reference point or standard called the "reference potential" to compare the voltages. In this way, one is able to say which reaction is more "noble" (*i.e.*, it does not take place easily) and which one is more "active" (thermodynamically easier to happen). Reference reactions are defined as having the following characteristics:

- All of the substances taking part in the electrode reactions must have unit activity (activity = 1).
- Temperature must be 25°C.
- Hydrogen pressure in the reference electrode must be one atmosphere.

By ranking and rating the electrode reactions according to the values of the standard potential, the *electrochemical series* is obtained. With respect to hydrogen voltage (accepted as zero), the potential required for reduction of Cu^{2+} to Cu is +0.34 volts and the potential of reduction of Fe^{2+} to Fe is –0.41 volts on the standard hydrogen electrode (SHE) scale. As reduction of copper ion to copper requires less energy (in terms of potential) than reduction of ferrous, then Cu^{2+} to Cu takes place more easily and sooner. The net effect is cathodic reaction of copper and *anodic* reaction of iron to ferrous resulting in corrosion of iron in copper sulphate solution, in other words, galvanic corrosion of iron by copper cathode ($Fe + CuSO_4 \rightarrow FeSO_4 + Cu$).

Although using the electrochemical series seems to be very useful, it does have very serious limitations such as [1]:

- Electrochemical series applies only to oxide-free metal surfaces and at the activities (concentrations) for which the standard potentials are valid. However, in actual practice, oxide films often cover the metal surfaces.
- The activities can deviate considerably from 1, especially when the metal ions are associated with other constituents in the so-called complex ions. It is not always possible to maintain standard temperatures and pressures.

Such conditions can result in the measured potentials having a completely different order than that given in the electrochemical series. That is why metals are exposed to a given electrolyte, say, seawater, and then arranged according to the measured electrode potential. What is then produced is called a *galvanic series* in a given environment at a given temperature. The potentials, however, are only valid for the electrolyte in which the measurements have been made. In seawater at 25°C, stainless steel (18/8) in the passive state has a potential of 0.19 V (measured with respect to hydrogen scale), which will make it more noble to galvanised steel that has a voltage value of –0.81 V (also with respect to hydrogen scale).

1.4 Corrosion Forecast

Pourbaix diagrams: Thermodynamically, one may forecast or predict whether certain reactions can occur; however, this does not mean that the reaction in question *will* occur. Thermodynamics can tell us about the "tendency" of corrosion to happen, and it is by using Pourbaix diagrams that one can thermodynamically predict whether certain reactions will occur.

Pourbaix diagrams use potential (in Volts) versus pH. A more simplified Pourbaix diagram is seen in Figure 1.2.

As seen in Figure 1.2, two lines designated by O_2 and H_2, upper and lower, define the domains for oxygen and hydrogen stability: above the upper line water is oxidised to O_2, so oxygen is evolved above the upper line. Below the lower line, water decomposes to H_2 and thus, hydrogen will be liberated below the hydrogen line. Hence, the domain with dark colour represents where corrosion occurs solely by oxygen reduction, and the domain with lighter colour represents where hydrogen evolution can also take place.

These diagrams can be used to follow the consequences of environmental changes upon corrosion behaviour. Pourbaix diagrams are useful guides as to "what should occur", but what determines "what actually does occur in practice" are the rates of the processes and reactions. To understand kinetics of corrosion, or in other words, to understand how fast corrosion takes place, one may use another tool that is known as polarisation curves. To understand polarisation curves, it is necessary to understand the their two components – anodic and cathodic polarisations.

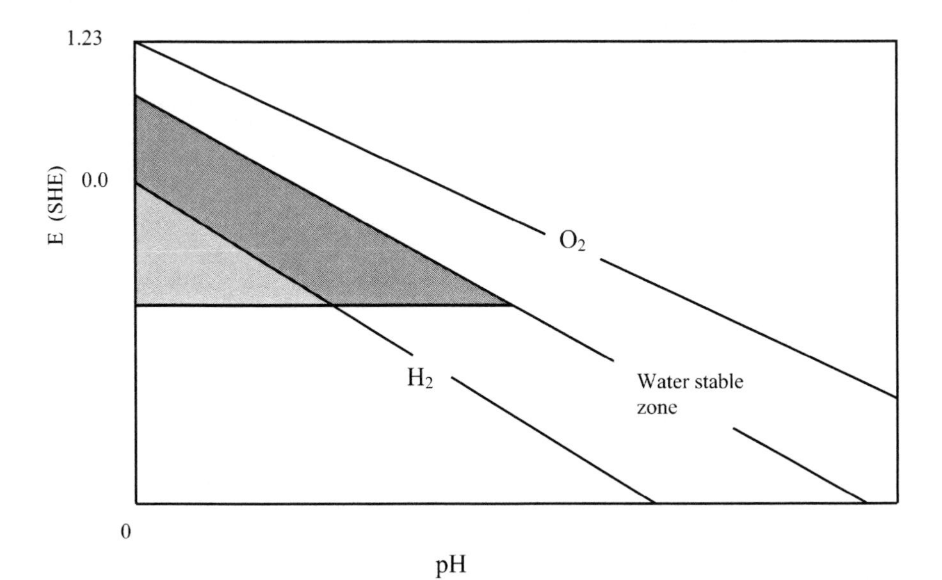

Figure 1.2 A simplified Pourbaix diagram

1.4.1 Anodic Polarisation (Anodic Polarisation Reaction)

An example of such polarisation curves is given in Figure 1.3.

Such curves are observed (a) in solutions with very low pH (sufficiently acidic) and (b) if the metal is capable of forming films. In this type of curves, three regions can be differentiated:

- The first region is where anodic current, and hence the dissolution rate, will increase as the anode potential is made more positive (high dissolution) ... (A → B)
- The second region is where film formation and passivation of the metal beneath happen. Formation of the film will act as a barrier to further dissolution so that the current (*i.e.*, corrosion rate) will fall ... (B → C)
- The third region is where the rise in potential (passive state) occurs ... (C → D)
- At very high potentials (point D) three possibilities, depending on the conductivity of the corrosion products film, can occur:
  - If the film has a good electrical conductivity, it may be oxidised to soluble species, and thus dissolution starts again along (D → E), where the metal transfers into the "trans-passive" state. This behaviour may be observed where continued dissolution of the passive film on chromium, Cr_2O_3[Cr(III)], to chromate, CrO_4^{2-}[Cr(VI)], takes place.
  - An alternative is if the film is a good electronic conductor, oxygen evolution may occur along (F → G).
  - However, if the film is a poor electronic conductor, then high anodic potentials may be reached along (F → H) with a constant, high-current density. This may facilitate "anodising", which is commercially used as a process for protecting aluminium.

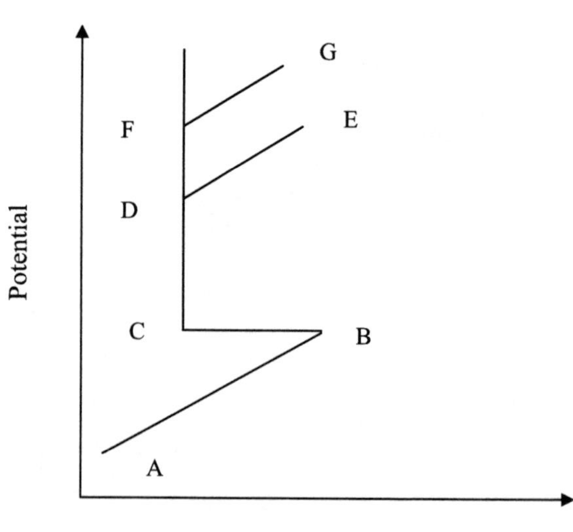

Figure 1.3 A typical anodic polarisation curve

It must be said that the presence of ions such as chloride ions may prevent or delay passivation. Furthermore, passive films – especially in stainless steels – will tend to break down in the presence of such ions.

1.4.2 Cathodic Polarisation (Cathodic Polarisation Reaction)

Examples of such polarisation curves are shown in Figure 1.4 as dashed lines.

In Figure 1.4, the cathodic reaction designated as 2 has a larger corrosion rate than cathodic reaction 1 ($I_2 > I_1$). With a cathodic reaction such as 3, passivation of the metal is reached. As mentioned earlier, there are a number of cathodic reactions such as hydrogen and/or oxygen evolution reactions, reduction of ions such as Fe^{3+} or MnO_4^- or molecular species such as nitric acid. In reality, the cathodic curves are not as straight as the lines shown in Figure 1.4. For example, the shape of the reduction curve of oxygen is rather "curvy", due mainly to the limited solubility of oxygen in aqueous solutions.

When a metal is exposed to an aqueous solution containing ions of that metal, on the surface of the metal, both *oxidation* (changing the metal atoms into the metal ions, or *anodic reaction*) and *reduction* (changing the metal ions into the metal atom, or *cathodic reaction*) can occur. When current is applied to the electrode surface, the electrode potential is changed, and it is said that the electrode

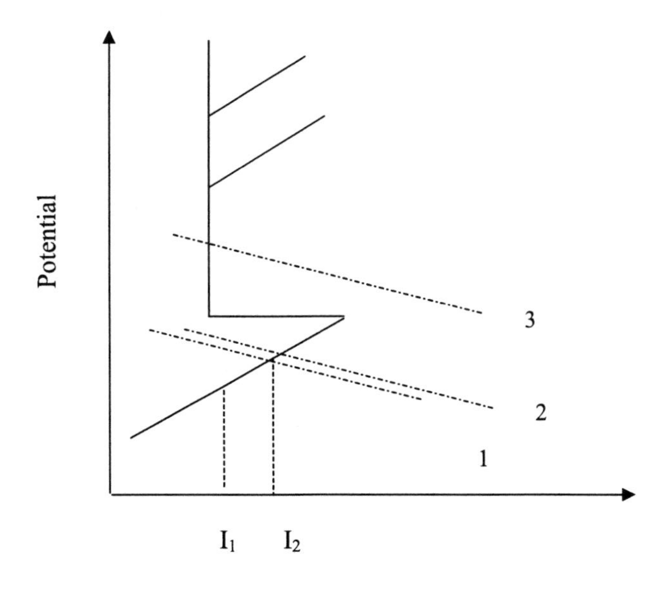

Figure 1.4 Cathodic polarisation where the dashed lines 1, 2, and 3 represent different cathodic reactions that, according to their potentials, can occur

has been *polarised* and the change in electrode potential is called *polarisation*. *Overpotential* is the difference between the electrode potential of the electrode with applied current and the equilibrium potential for the electrode reaction in question. *Depolarisation* is then defined as the removal of factors diminishing the rate of an electrochemical reaction, for example, cathodic reaction.

1.5 Summary and Conclusions

In this chapter, we very briefly touched on some important theoretical elements of electrochemical corrosion such as how corrosion can be forecast (by using Pourbaix diagrams) and how fast it can happen (by using anodic and cathodic polarisation curves).

Pros and cons of the application of the polarisation methods in studies related to microbial corrosion are addressed in Chapter 6. In the next chapter, the implementation of theoretical electrochemistry will be discussed to show how and why some techniques such as inhibitor addition or coating application work in practice.

References

1. Mattson E (1989) Basic corrosion technology for scientists and engineers. Ch. 3. Ellis Horwood Publishers, West Sussex, England
2. Videla HA (1996) Manual of biocorrosion. Ch. 4. CRC Press, London
3. West JM (1986) Basic corrosion and oxidation. Ch. 6. Ellis Horwood Publishers, West Sussex, England

Further Reading

Fontana MG (1987) Corrosion engineering. 3rd edn. McGraw-Hill International Editions
Scully JC (1983) The fundamentals of corrosion 2nd edn. Pergamon Press, UK
Shreir LL, Jarman RA, Burstein GT (eds.) (1995) Corrosion. 3rd edn Butterworth-Heinemann Ltd., Oxford, UK

Chapter 2
Technical Mitigation of Corrosion: Corrosion Management

2.1 Introduction

Corrosion can be mitigated by two approaches. One method, known as the "technical approach" includes all known mitigation techniques such as design and application of cathodic protection, using inhibitors and the like. In the literature of corrosion, this approach has another name: corrosion management. In this chapter, we will discuss "corrosion management".

2.1.1 Corrosion Management: A Technical Approach

In this section, some of the technical methods which are frequently used to solve the problem of corrosion and reduce its effects are discussed. These methods, in principle, may include:

- Coatings and linings
- Anodic and cathodic protection
- Use of inhibitors
- Material selection and design improvement

These methods will be discussed in the sections below. Care will be taken not to go through all the details of each technique but rather link the technique with the theoretical aspects of electrochemistry, as covered in Chapter 1.

2.2 The Rationale for Using Coatings and Linings

The main benefit of using coatings and linings is that they prevent the electrolyte from coming into contact with the electrodes. In this way, there will not be an interaction between the anode and the cathode. The "electrochemical triangle", from Chapter 1, explains how and why, by using coatings and linings, the sides of the triangle are broken and thus no electrochemical corrosion may be expected.

2.3 The Rationale of Anodic and Cathodic Protection

2.3.1 *Anodic Protection*

Referring to Figure 1.3 in Chapter 1, the passivation of metal occurs within the passive range (C $\rightarrow$ D) so that if the potential is kept that high, the metal will anodically be protected. However, anodic protection has some drawbacks, such as:

- The metallic structure must be of a material of suitable composition for passivation in the particular solution.
- If protection breaks down at any point, corrosion will be extremely rapid at that point because of the low resistance path formed. On the other hand, if the metal potential is made too positive, then the region of passivation may be passed and transpassive corrosion in the form of pitting will occur. To avoid such things happening, extensive monitoring and control facilities are required.
- The passive films will be destroyed if aggressive ions such as chloride ion are present.

Anodic protection is achieved by applying an external cathode and a counter electrode in a manner similar to cathodic protection (see the next section), except that the current direction is in the opposite sense.

Anodic protection has found some applications in the fertiliser production industry to control corrosion of mild steel in contact with ammonia-ammonium nitrate solutions. It has also been used in vessels containing sulphuric acid.

2.3.2 *Cathodic Protection (CP)*

The corrosion rate of a metal surface in contact with an electrolyte solution is strongly dependent on the electrode potential. In most cases, the corrosion rate can be reduced considerably by shifting the electrode potential to lower values, such as those shown in Figure 2.1.

The main target here is to reduce corrosion by lowering the potential either (1) by connection to an external anode (sacrificial anode) which is a metal more active than the corroding metal, or (2) by adjusting the potential of the material by application of an external current (impressed current). Referring to Figure 2.1, it is seen that by reducing the potential of the metal from E_1 to E_2 and further down to E_3, the corrosion current is reduced (note how I_1 is reduced to I_2). It follows that

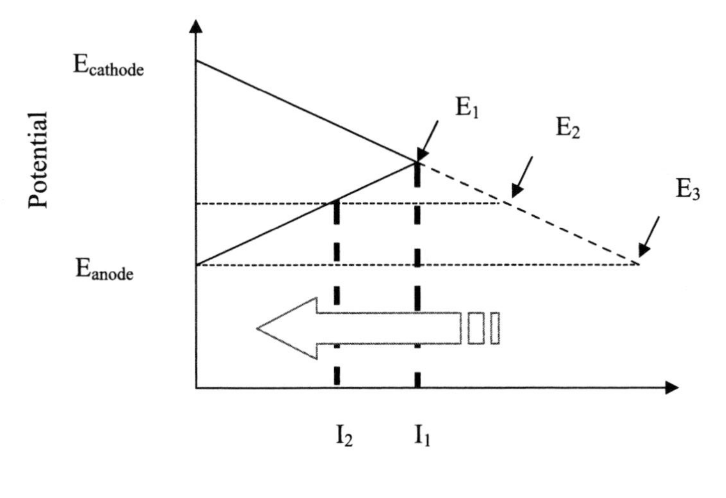

Figure 2.1 Principle of cathodic protection. At a potential like E_1, a corrosion current (rate) such as I_1 is reached. By making the potential more negative to E_2, the corrosion current is reduced to I_2, as shown by the red arrow

if the potential is further reduced to E_{anode}, then the corrosion current will be zero and the cathodic protection will be completed.

The main concern in CP is that there will be an increase in the alkalinity of the environment produced by the cathodic reaction (see Equation 1.3). This is important because many metals such as iron, aluminium, and zinc are affected under high pH conditions. If paints have been used with a CP system, they must also withstand the alkalinity of the medium. The basic criteria for CP, using an Ag/AgCl seawater reference electrode for the potential measurement, is a negative voltage of at least –0.80 V between the reference electrode and the structure.[1]

Some points about sacrificial anodes and impressed current CP systems in offshore structures are: [1]

- A widely used anode in impressed current CP systems for offshore structures has been lead-6%, antimony-1% silver alloy. Other anode materials that have been used with some success are lead-platinum, graphite, and a silicon-iron-chromium alloy. The lead-antimony-silver anodes may either be suspended or placed in special holders for rigid attachment to the underwater platform members. Suspended systems are somewhat more susceptible to mechanical damage, but they are simple to install and relatively easy to maintain. Impressed current systems are capable of long-term protection but are less tolerant of design, installation, and maintenance shortcomings than sacrificial anode systems. Routine comprehensive system monitoring is a must.

[1] More on this subject will be explained in Chapter 9.

- Alloys for offshore platforms may be alloys of magnesium, zinc, or aluminium. Different methods may be used to attach the anodes to the structure depending on their type and application, but most importantly, a low-resistance electrical contact must be maintained throughout the operating life of the anodes. Most sacrificial anode CP systems installed on new structures utilise aluminium alloy anodes. It is easier to design and control the current density with sacrificial anode (galvanic) systems than impressed current systems. Anodes are selected to provide a specific life, often 20 to 25 years. It is important to note that the total weight of all anodes must be included in the structural design calculations for the platform. Otherwise, practical problems in selecting a CP system may arise.

2.4 Use of Inhibitors

Inhibitors can be classified as anodic and cathodic. Each of these types of anodes has its own properties depending on the way it affects anodic and cathodic reactions. However, it must be noted that inhibitors are mainly used to affect the electrochemistry of the system and not its microbiology. Below, brief explanation of these types of inhibitors are given [2–4].

2.4.1 *Anodic Inhibitors*

These types of inhibitors can control the rate of oxidation (anodic) reactions. Anodic inhibitors are of wide variety and include chromates (CrO_4^{2-}) and nitrites

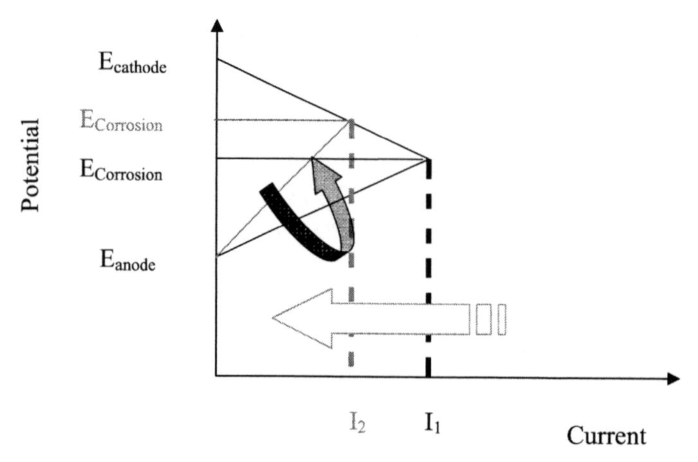

Figure 2.2 Principle of anodic inhibition. When no anodic inhibitor has been added, the corrosion current (rate) is I_1. When an anodic inhibitor is added, the anodic potential is modified and "rotates up" so that the corrosion rate is reduced (I_2)

(NO_2^-), which are oxidising anions, and other compounds such as silicates, phosphates, benzoates and molybdates which are non-oxidising, acting in neutral or alkaline solutions. The mechanism is that the anodic reactions become highly polarised and the mixed corrosion potential of the specimen under such conditions is shifted in the noble direction as shown in Figure 2.2.

In the presence of air, certain anodic inhibitors such as phosphate and molybdate form a protective (passivating) oxide layer on the metal surface. If the inhibitor concentration is too low, pores and defects can arise in the oxide layer, where accelerated corrosion can take place. These inhibitors are therefore called "dangerous inhibitors" [2].

2.4.2 Cathodic Inhibitors

These types of inhibitors can prevent or reduce the rate of reduction (cathodic) reactions. When these inhibitors are used, the mixed potential is lowered and there is again a decrease in corrosion current, as shown in Figure 2.3. Examples of cathodic inhibitors are [2]:

- Zinc salts, *e.g.*, $ZnSO_4$; their action depends on zinc hydroxide being precipitated at the cathode, where the pH increases, thus making the cathode reaction more difficult.
- Polyphosphates, *e.g.*, sodium pyrophosphate ($Na_4P_2O_7$), sodium tripolyphosphate ($Na_5P_3O_{10}$), and sodium hexametaphosphate ($(NaPO_3)_6$) that in the presence of divalent metal ions form a protective coating on the metal surface.

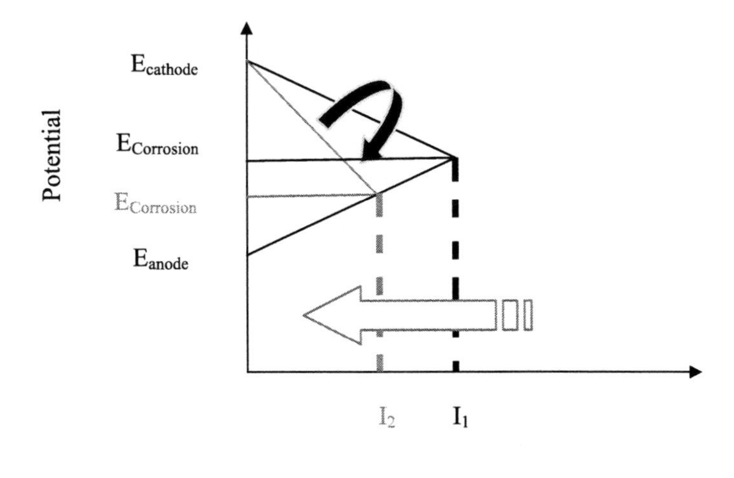

Figure 2.3 Principle of cathodic inhibition. When no cathodic inhibitor has been added, the corrosion current (rate) is I_1. When a cathodic inhibitor is added, the cathodic potential is modified and "rotates down" so that the corrosion rate is reduced (I_2)

- Phosphonates, which in the presence of two-valent metal ions and preferably in combination with a zinc salt are effective as inhibitors.

Even with a low concentration, cathodic inhibitors provide some inhibition (in contrast to anodic inhibitors). Therefore, they are not "dangerous" at concentrations which are too low for complete inhibition.

2.4.3 *Mixed Effect Inhibitors [2]*

Some inhibitors function as both anodic and cathodic, influencing both the anode and the cathode reactions to a larger or lesser extent.

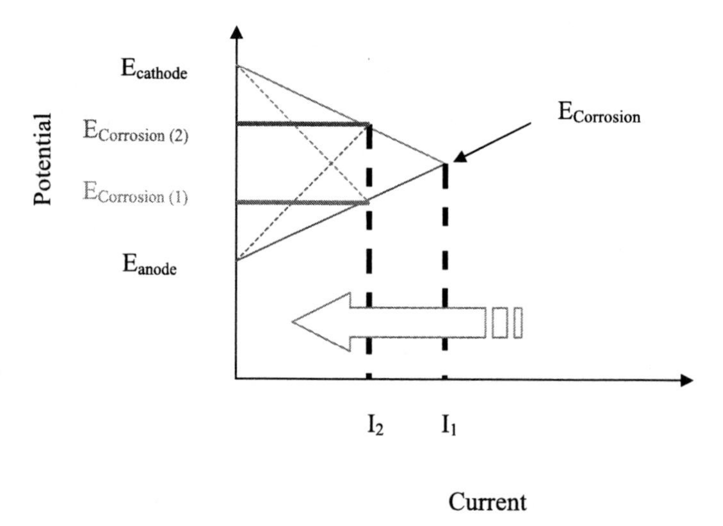

Figure 2.4 The mixed effect of inhibitors, affecting both anodic and cathodic reactions to reduce corrosion

As an example of these types of inhibitors, polyphosphates, phosphates, silicates, and benzotriazole can be mentioned. The action of these inhibitors is highly dependent on the environmental factors such as pH and redox potential. Therefore, they are anodic under certain conditions and cathodic otherwise.

2.5 Material Selection and Design Improvement

One of important challenges in combating corrosion is replacing available material with an upgraded, more resistant material. Frequently, a material is selected based on its resistance to corrosion, but this resistance is a result not only of the physical and chemical features of the material, but also its working conditions. In Chapter 8, some important materials are described which are not immune to microbial corrosion, despite some myths that have surrounded them regarding their resistance to this type of corrosion.

Most of the time, the upgraded material is not economically compatible with the existing one and the cost of removing old material and replacing it with the new one will also add up more to the question of the feasibility of the program. Therefore, the design engineer must pay attention to a range of factors including the intrinsic resistance/vulnerability of the material to the service conditions as well as the environment in which the material is put into the service and the economy of using a certain material in a given environment. For instance, sometimes, stainless steel 304 can be as vulnerable as carbon steel with regard to microbial corrosion. However, despite all the cost that using corrosion resistant materials may impose on the financial framework of a project by increasing the costs at the design stage, by considering the losses (especially economical losses) resulting from corrosion, in the long-term using corrosion resistant materials will be justified and beneficial.

Corrosion prevention by design modification may seem too simple at first glance, but it is not applied in many designs. The following are just some general guide lines for designers to help them avoid corrosion and especially microbial corrosion problems in their designs:

- Avoid designs that allow for water or dirt collection/stagnant water and/or moist accumulation. If the system becomes too dirty because of being suitable for dirt and debris collection, the environment may become very receptive to microbial species capable of affecting both the extent and intensity of corrosion. In pipelines, for instance, designs that allow too many ramifications and extra branches and piping can render the pipe quite vulnerable to microbial corrosion.
- Avoid designs that cause turbulence, such designs may indirectly help microbial corrosion by promoting the possibility of erosion-corrosion and thus producing an environment which is already corroding and helping bacteria with the required ferrous ions for example.

In Chapter 5 some important factors contributing to rendering a system vulnerable to microbial corrosion have been explained. Avoiding having such factors in the system can assist in safe-guarding against microbial corrosion. In Chapter 9, treatment of microbial corrosion in more detail will be explained.

2.6 Summary and Conclusions

Corrosion management deals with the study and implementation of techniques and methods such as cathodic protection, coatings and materials selection to mitigate corrosion, and of course, microbial corrosion. In this chapter, the main basics and logics of some of these important techniques were explained. The next chapter deals with a very new concept called corrosion knowledge management which more than concentrating on the technical methods and technologies, relies on team-building and managerial aspects of managing corrosion.

References

1. Byars HG (1999) Corrosion Control in Petroleum Production. TPC Publication 5, 2nd edn. NACE International, USA
2. Mattson E (1989) Basic corrosion technology for scientists and engineers. Ch. 3. Ellis Horwood Publishers
3. Miller JDA and Tiller AK (1970) Microbial aspects of metallurgy. Miller JDA (ed.), American Elsevier Publishing, New York
4. West JM (1986) Basic corrosion and oxidation. Ch. 6. Ellis Horwood Publishers

Chapter 3
Non-technical Mitigation of Corrosion: Corrosion Knowledge Management

3.1 Introduction

Corrosion problems can be approached from different, yet not necessarily opposite, points of view. Solving a corrosion problem as a technical problem requires applying corrosion mitigation and prevention-related technologies in the domain of corrosion management, as described briefly in Chapter 2.[1] There is, however, another dimension to this, and it is the way that a corrosion engineer has to communicate with the mangers who may not have the same level of expertise in corrosion management. Here, the problem is not a technical problem any longer; it is related to using a language which, while corrosion-related, is not that technical. At the same time, this language must be something that will address human relationships and its impact on the management. This is what we may call "corrosion knowledge management".

In other words, corrosion engineers can have highly sophisticated knowledge of what is to be involved in the mitigation and prevention of corrosion. This will give them a language of their own, a language that is highly technical and may sound like "jargon" to others. On the other hand, the managerial level of a plant or industry has to deal with many issues, one of which is corrosion. This will create a language and "culture" of its own so out of touch for a technical person that

[1] The author "learnt" to emphasise upon the difference between corrosion management (CM) and corrosion knowledge management (CKM) when he realised that the management-based principles of what he had called as CM were understood by corrosion community as to be not different from technical context of the so-called "corrosion management". To highlight that this technique is entirely different from the known, so-called corrosion management concept, the author had to make use of the suffix "knowledge management" so that the managerial side of this approach is preserved. However, my rather "early" works still have the label of "corrosion management". In tis context, then, whatever we refer to as corrosion management (CM) must be understood as corrosion knowledge (continued) management (CKM). The author apologises in advance for any confusion and inconvenience thereafter that may be created.

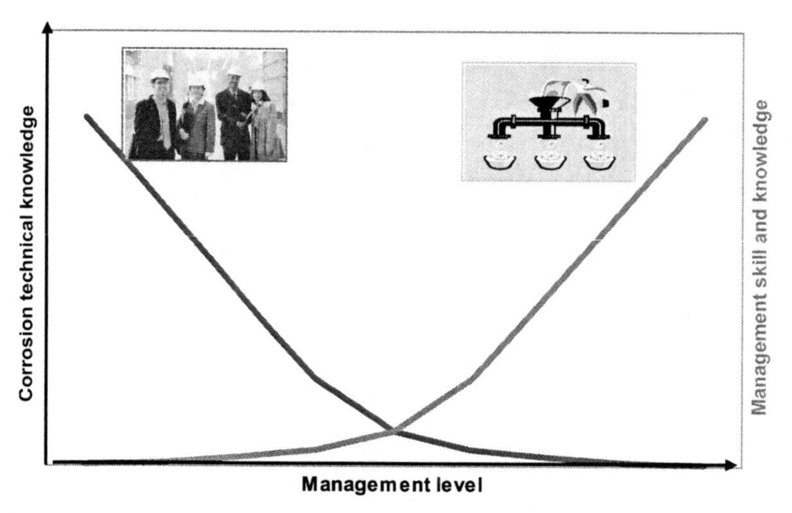

Figure 3.1 How can corrosion engineers, with a high level of corrosion technical knowledge, and managers, who must think of making the industry more profitable, reach a common language?

understanding managerial decisions (on either micro- or macro-management levels) may not always make sense (Figure 3.1).

In this section, we will explore how such a gap can be filled by a systematic approach towards the problem of corrosion that will use no "corrosion management" language.

3.2 Corrosion (Knowledge) Management: A Managerial Approach

Here we describe some tools with which managers with no hands-on experience in corrosion may become confident about the best strategy to take against corrosion. The main point here will be advising managers to arrange their resources in a more feasible way to mitigate corrosion. Corrosion Knowledge Management may be defined as the shortest and least expensive way to control corrosion in terms of resource management.

3.2.1 Importance of Defining "Resources" and "Targets"

Among definitions that may be found for management, what one can intuitively recognise is that management is, indeed, an "art and science" of balancing between what you have in hand and what you want to achieve by using the more practical, the least expensive, and the shortest path (route). In other words, a good manager is the one who:

- Knows the resources (R)
- Understands the targets (T)
- selects the path (P) to reach (T) via (R)

A good manager should also try to find answers for the following questions:

- **W**hat are my resources?
- **W**hy do I choose certain target(s)?
- **H**ow should I get to the target(s) by using the resources?

The above questions may be referred to as "2W-H" questions. The first **W** deals with what a manager has got in hand, *i.e*, **R**. This can include many factors such as human resources, financial resources, *etc.*, but in addition to those, a manager as a professional is also considered an important resource. For this reason a good manager must have:

- Knowledge and
- Information

Knowledge in this context means the minimum qualifications one has about a topic, in other words, having an academic degree as a minimum. The knowledge gained by getting a degree gives the ability to:

- Define the present state of a system
- Predict the future state of the system

However, one has to update one's knowledge. The continuous process of renewal and updating one's knowledge can be called "information". Any processed data entering into the territory of our knowledge to improve it according to existing conditions is called information. To know the resources, a good manager *must* have both knowledge and information about them.

So for a manager, information means a continuous process of being aware of everything related to his/her specific job: if he is a manager of an art museum with a B.Sc. in mechanical engineering, he would definitely have to study more about his particular job and its new developments than a manager with a degree in art.[2] Similarly, an engineer who has no degree/skills in corrosion but is committed to manage a plant has to be more aware of corrosion to be able to take a holistic approach to protect his assets.

Why should a manager make a choice on achieving a particular target? Why should he/she select this, and only this particular one, as the goal to achieve? There are many factors that dictate how and why to choose certain things as goals or targets. Some of these factors are:

[2] It is a fact of life that managers may not always necessarily have been technically trained for their positions.

- Social reasons
- Political reasons
- Economical reasons
- Cultural reasons
- A combination of the above

To make it more clear, an example of building a power plant may be useful: some of the goals (targets) for such a project can be economic (such as solving the problems related to the lack of electricity in the region.), political-economical (winning the elections, to create new jobs and work opportunities ...), *etc.* Therefore, based on which factor becomes more relevant, a certain target may be taken or left.

To answer to the question "HOW" of the 2W-H questions, one has to consider the "work scheduling", three principals of which are:

- Knowing elements of the process/project and their relationships
- Preparing executive time schedule
- Estimation of expenses and required budget

It must be said that of the three principles mentioned above, the first is the one that makes a project feasible or not. The other two factors are mechanically calculated according to the "complexity" of the project in terms of factors involved. We will see the importance of knowing elements of a project when we are introducing the concepts of *corrosion of a system* and *corrosion in a system.*

3.2.2 Why Should We Care About Corrosion?

For a manager, the reason to fight corrosion is to reduce the losses. As corrosion-related losses are very important, we will consider them under a separate section here; however, it must be emphasised that this section is still linked to the topic of the importance of defining targets.

Losses due to corrosion can be divided into three categories [1]:

1. Waste of Energy and Materials
2. Economic Loss
   a) Direct Loss
   b) Indirect Loss
      - Shutdown
      - Loss of efficiency
      - Product contamination
      - Overdesign
3. Environmental Impact/Health

We will discuss each of these categories briefly.

Waste of Energy and Materials

Internationally, one ton of steel turns into rust every 90 seconds. On the other hand, the energy required to make one ton of steel is approximately equal to the energy an average family consumes over three months [2]. As another example, take a pipe line of 8-in. diameter and 225 miles (~362 km) long and a wall thickness of 0.322 inches; with adequate corrosion protection wall thickness could have been only 0.250 inches, thus saving 3700 tons of steel as well as increasing internal capacity by 5% [1]. Of every ton of steel from the world production, approximately 50% is required to replace rusted steel [2]. Reports show that the loss of a Sea Harrier in the Adriatic in December 1994 and partial structure loss in a 19-year-old Aloha Airlines Boeing 737 in April 1988 can both be attributed to **corrosion** [3].

Economic Loss

When economy is mentioned, it can be looked at as both micro- and macro-models. In other words, one can calculate both how corrosion takes away from the pocket of the man on the street and how the national economy is affected by corrosion. The figures here are focused more on national and macro-scale economy figures but, as the reader will also appreciate, at the end of the day, it is the man on the street who pays.

Insurance companies have paid out more than US$ 91 billion in losses from weather-related natural disasters in the 1990s [4], whereas direct loss of corrosion in 1994 just in the U.S. industry was US$ 300 billion [5]. The cost of corrosion has been reported from many studies to be of the order of 4% of the GNP (Gross National Product) of any industrialised country [6]. In the power industry, it has been estimated [7] that corrosion losses in utility steam systems amounted to about US$ 1.5 billion of the US$ 70 billion annual cost of corrosion in the U.S. in 1978.

Figures show that corrosion costs the U.S. electric power industry as much as US$ 10 billion dollars each year [8]. New studies on updating U.S. corrosion costs in 2000 have shown that the total cost of corrosion in 1999 dollars to remediate corrosion-induced structural deficiencies of highway bridges was estimated at approximately US$ 30 billion [9]. The same study showed that the current cost of corrosion protection built into new automobiles determined by auto manufacturers and other experts is US$ 150 per vehicle. In fact, the percentage of the GNP attributed to motor vehicle corrosion in 1998 was 0.25%. Other sources [10] report that BP had performed pigging on its Prudhoe bay pipelines more than 350 times in 2005. An example of the global "ripple" caused by this disaster was that when BP said that it would stop the flow of half as much oil in the summer of 2006, the price of oil increased by 3.4 percent, skyrocketing to US$ 77.30 a barrel the next day [11].

Environmental Impact/Health

Table 3.1 shows some examples of recorded environmental/health impacts of corrosion in different years, countries, and places [12, 13]: Shipilov recently re-

Table 3.1 Some Corrosion-Related Incidents

Year	Place	Accident	Probable reason	Results
1970	North Sea	Platform collapse	Stress corrosion cracking (SCC)	Huge life, material and economic loss
1967	Ohio River (U.S.)	Collapse of the "Silver Bridge"	SCC	Huge life and material loss
1985	Switzerland	Collapse of the 200-ton concrete ceiling of an indoor swimming pool	SCC in stainless steel bars holding the ceiling, due to existing chlorine ions	12 people died, others were injured
1996	Mexico	Fire and explosion	Petrol leaking from a valve on a 1,300 m^3 storage tank caught fire, causing the tank to explode	Four people died and 16 were injured. The Red Cross tended to 960 people and 10,000 were evacuated. It took two days to bring the fire under control
1997	Canada	Spill of over 35,000 litres of oil in one night	A leak in a damaged pipeline owned by Mobiloil	Large scale environmental pollution
1997	Russia	Leakage of over 1,200 tons of oil	Leakage from a ruptured pipeline	About 400 tons of oil spilled into the River Volga. A dam was built in a tributary of the river to prevent further pollution

viewed economic and environmental impacts of corrosion all around the world [14]. The 2006 accident at Prudhoe Bay resulted in an environmental/hazard nightmare: reports indicated that[3] the leaking crude oil formed a black layer over an area of grassland about the size of half a football pitch, where, according to BP, the clean-up cost will be over US$ 200 million.

Such studies show that the impact of corrosion on the environment and health and safety is so large that ignoring it can cause very serious consequences. Therefore, corrosion and its environmental impacts have the potential of producing a very wide range of hazards and disasters. Rephrasing Shipilov's words, corrosion is indeed "history's worst silent serial killer".

3.3 Components of Corrosion Knowledge Management as a Managerial Tool

Corrosion knowledge management (CKM), in essence, doesn't differ from other management approaches. CKM requires one to consider "R","T", and "P" as well

[3] THE TIMES, Saturday, 12/August/2006. The author would like to thank Dr. Roger King for providing him with these newts. It is "rumoured" that the cause of the failure could be microbial corrosion.

as answering the 2W-H questions in order to decrease unwanted effects of corrosion. CKM in fact summarises what industry can eventually end up with in dealing with corrosion in a more ordered, systematic way.
R (Resources) in CM are:

- Capital
- Expert or expert team
- Training
- Research for new anti-corrosion materials
- Research for new methods used to control corrosion
- Information
- Energy
- Time

T (Target) in CKM is:

- To control corrosion to lower its costs (economic-ecological reason)

P (Path) in CKM is, then:

- Corrosion Knowledge Management

The above means that to deal with corrosion and decrease its costs efficiently, managers must reconsider their "R" and "T" according to principles of CKM. In other words, if they see there is a corrosion problem in their systems, first they have to check their resources to see if there are enough to mitigate the problem. It should be noted, however, that setting targets (the Why question of the 2W-H questions) would determine which factor is missing in resources or which factor is worth more consideration, and so on.

CKM has four components [15]:

1. Modelling
2. Use of information
3. Transparency
4. Corrosion system definition (COFS/CINS)

The first three items have been discussed to some extent in the related literature somewhere else [16, 17, 18]. However, it is worth it to concentrate more on the last item, *i.e.*, corrosion system definition.

A very important aspect of solving corrosion problems is understanding the system in which corrosion is taking place. A corrosion system is defined as ***part of a universe in which corrosion occurs and is of interest to us***. Corrosion system can be considered as consisting of subsystems such as A, B, C, *etc.* If a corrosion problem of each subsystem "i" is shown as corr(i) and corrosion types observed in each subsystem as a, b, c, …, one can write:

$$\text{Corr(A)} = \{a_1, a_2, a_3, \ldots, a_n\}$$

$$\text{Corr(B)} = \{b_1, b_2, b_3, \ldots, b_n\}$$

$\mathrm{Corr}(C) = \{c_1, c_2, c_3, \ldots, c_n\}$

Corrosion of a system, or ***COFS*** is then defined as:

$\mathrm{COFS} = \mathrm{Corr}(A) \cup \mathrm{Corr}(B) \cup \mathrm{Corr}(C)$

Corrosion in a system, or ***CINS***, is then defined as:

$\mathrm{CINS} = \mathrm{Corr}(i) \cap \mathrm{COFS}$

As an example, take corrosion in an automobile: suppose we define a car as a corrosion system – that is to say typical types of corrosion in subsystems of a car. In this case, subsystems can be defined as:

A = Chassis, B = Fuel system, C = brake system, ...

Corrosion problems in each of the above subsystems, with important mechanisms in parenthesis, can be shown as the following [18]:

Corr(A) = {uniform, pitting, (crevice), fretting, stress cracking, ...}

Corr(B) = {(pitting), crevice, coating failure}

Corr(C) = {(pitting), crevice, galvanic, (fretting), (coating failure)}

Thus a project whose goal is to solve *all* corrosion problems of a car would have to deal with *all* of the corrosion problems in *all* subsystems, in other words, it would be a COFS approach. In this case, study of the corrosion of just a given subsystem, such as Corr(C), will be a CINS approach.

It is very important to distinguish between CINS and COFS approaches, because many problems such as the expected time span of the project or required capital for doing the project may vary greatly. A practically important alternative definition of corrosion is a system with "highest risk". More often than not, a large percentage of the risk (> 80%) is found to be associated with a small percentage of the equipment item (< 20%) [19]. Once identified, the higher-risk equipment becomes the focus of the inspection and maintenance to reduce the risk, while opportunities may be found to reduce inspection and maintenance of the lower-risk equipment without significantly increasing risk. In other words, to be on the safe side, it is better to choose the system of concern – the one with higher risk – and define COFS & CINS according to real, working conditions of the system.

CKM is schematically represented in Figure 3.2.

In dealing with corrosion, a manager needs to consider why corrosion mitigation is important (or not important). If he decides to cure a corrosion problem, for one or more of the reasons we mentioned earlier in this section, then he should check his resources. Some items in the resource box may not be as urgent as others; for example, the manager may not be that concerned about the outcome of research on new and rather "exotic" corrosion-resistant materials and/or detection devices as his immediate concern could be solving a problem that has the potential of doing damage and ruining his business. However, this manager must learn to work closely with researchers and research bodies to ensure that he is constantly

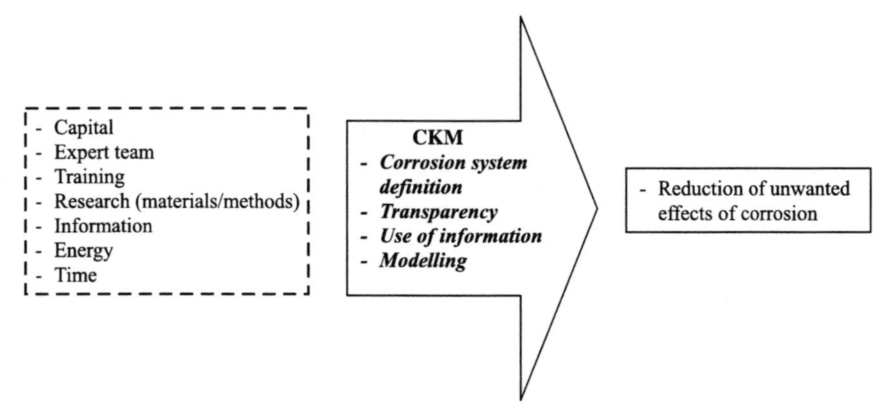

Figure 3.2 The relationship between the resources, the target, and the role of CKM

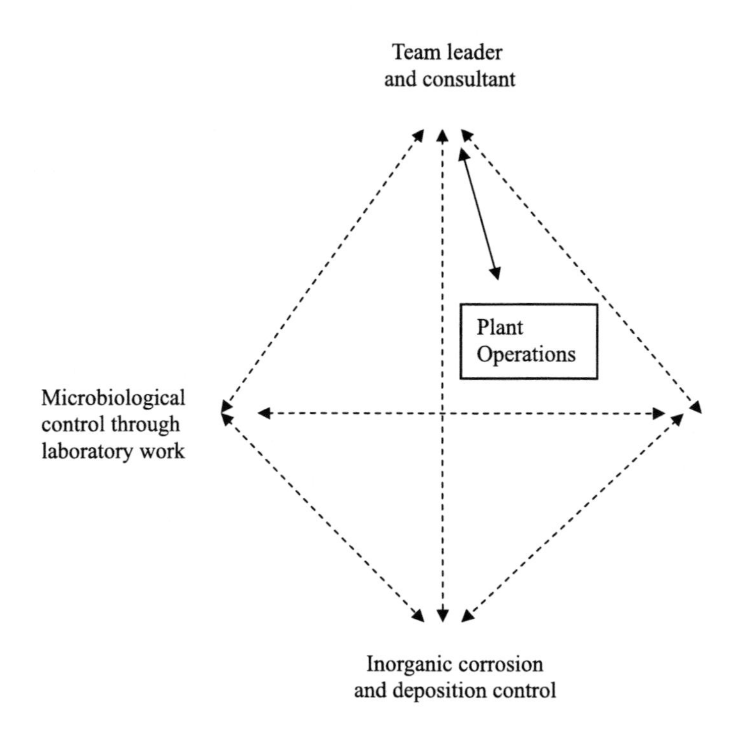

Figure 3.3 A model for communication between expert teams and consultants to deal with a complex corrosion problem

being updated. In fact, the larger the area of operation in the company, the more important the need for continued attention to and financial support for research.

Another important item is training. It may sometimes be economical and feasible to train the existing staff with regard to a certain discipline; for example, enrolling them in a cathodic protection system. However, depending on the severity of the problem, it may sometimes be necessary to recruit someone to build an expert team to deal with corrosion. Such a team may act in different ways to affect the severity of corrosion. A possible way to do so could be through by a communication network as shown in Figure 3.3.

Figure 3.3 illustrates an imaginary case of corrosion treatment that includes both microbial and non-microbial components. As seen in the figure, although an expert team is in direct communication with both the plant operation manager and the consultants for assessing non-microbial components of the problem, an advisory team is also involved and would comment on microbial corrosion and the findings obtained by an assigned laboratory. The expert team could be very good at dealing with non-microbial cases, but as experts, they would be prudent to realise that and seek technical advice from another group of experts who may know microbial corrosion much better.

Figure 3.3 also reveals that a CKM management unit must be supervising the activities requiring a CKM approach. Such a management unit must be able to assess the resources in accord with CKM principles and prepare the necessary feedback for the management to deal with corrosion more efficiently.

3.4 Conclusion and Summary

Corrosion knowledge management (CKM) is different from corrosion management. CKM can be regarded as an interface through which managers and engineers can communicate more effectively. This chapter explained briefly the principles of CKM.

References

1. Uhlig HH (1971) Corrosion and corrosion control. 2nd edn. John Wiley & Sons, West Sussex, England
2. New prefabricated joist designs researched (1997) Mater Perform 36(7):68
3. Hill KWM (1998) Corrosion prevention in the Royal Air Force: There is no option. Corrosion Management 26:14–19
4. Worldwatch News Brief 99-3, Destructive storms drive insurance losses up, www.worldwatch.ord, March 26, 1999
5. Cost of corrosion: $ 300 billion a year. (1995) Mater Perform 34(6):5

6. Heitz E (1992) A Working party report on microbiological degradation of materials and methods of protection. In: Heitz E, Mercer AD, Sand W, Tiller AK (eds) The Institute of Materials, England
7. Gaona-Tiburcio G, Almeraya-Calderon F, Martinez-Villafane A, Baustista-Margulis R (2001) Stress corrosion cracking behaviour of precipitation hardened stainless steels in high purity water environments. Anti-corrosion Methods & Materials 48(1):37–46
8. Bacteria could help control corrosion at power plants (1998) Mater Perform 37(11):50–51
9. Cost of corrosion study update: Trends in the automotive industry (2000) Mater Perform 39(8):104–105
10. Bailey A (2006) BP: Learning from oil spill lessons. Petroleum News 11(20). http://www.petroleumnews.com
11. BP Pipeline Failure Follow-Up (2007) Corrosion & Materials 32(2):7
12. Javaherdashti R (2000) How corrosion affects industry and life. Anti-corrosion Methods & Materials 47(1):30–34
13. Industrial accidents (1997) UNEP Industry and Environment 20(3):6
14. Shipilov SA (2007) University education in corrosion: A true challenge for the engineering world. Proceedings of The Iranian Corrosion/2007; ICA International Congress, May 14–17 2007, Tehran, Iran
15. Javaherdashti R (2003) Managing corrosion by corrosion management: A guide for industry managers. Corrosion Reviews 21(4):311–325
16. Javaherdashti R (2000) Corrosion Management: CM. www.nrcan.gc.ca/picon/conference2; Natural Resources Canada, December 18
17. Javaherdashti R (2002) How to manage corrosion control without a corrosion background. Mater Perform 41(3):30–32
18. Corrosion in Automotives. In: Baboian R (ed) Corrosion tests and standards manual: Application and interpretation, ASTM manual series: MNL 20 ASTM 1995
19. Hovarth RJ (1998) The role of the corrosion engineer in the development and application of risk-based inspection for plant equipment. Mater Perform 37(7):70–75

Chapter 4
Microbiologically Influenced Corrosion (MIC)

4.1 Introduction

One type of corrosion that can be very harmful to almost all engineering materials is what is called microbiologically influenced corrosion, or MIC.[1] The term MIC may be misleading, giving the impression that it is only micro-organisms that are capable of influencing corrosion. In fact, biofouling, a more general term, can be used to study both the microbiological and macrobiological growths that happen on surfaces and can show both enhancing and inhibiting effects [1].

MIC and the way it affects corrosion have always been a matter of debate. For example, while acid production by bacteria is presumed to be one of the ways by which corrosion can be enhanced, some researchers [2][2] in their experience with aerobic *Pseudomonas* sp. have reported that acid production was not a major cause of corrosion, and others [3][3] have pointed out that the presence of bacteria was not "an important factor in the deterioration of steels". It seems that it is not always easy to come up with a clear, once-forever-true explanation of the impact of bacteria on corrosion. As a matter of fact, such relatively confusing outcomes have helped make MIC a puzzle to some and to others an "industrial joke" that is used when there is no other explanation for the failure.

---

[1] In 1990, NACE officially accepted the term "Microbiologically Influenced Corrosion" to address this type of corrosion (see: Materials Performance (MP), September 1990, p. 45). This type of corrosion is also called "microbiologically induced corrosion", microbial corrosion or biocorrosion. In this book, all of these terminologies will be used interchangeably.

[2] While those authors have ruled out the effect of the acid produced by the bacteria on corrosion acceleration, they have suggested that in the presence of an aerobic hetertrophic bacterium, repassivation of pits does not happen but pit growth continues. They nominate pit propagation in the presence of bacteria as the main mechanism for observing the drop in carbon steel's open circuit potential (OCP) and polarisation resistance.

[3] These researchers reported, however, that the biofilm formed by the bacteria in their study could have a protecting rather than a deteriorating effect.

This chapter will deal with MIC, its definition and importance, and how historically both our understanding of and research methods for the study of MIC have evolved. We will then have a look at the parameters that can be used for categorising bacteria, and also the steps involved in biofilm formation. After discussing the ways by which biofilms can both accelerate and decelerate corrosion, we will look at three examples of bacteria that are involved in corrosion, the well-known SRB (sulphate-reducing bacteria), the rather "shy", infamous IRB (iron-reducing bacteria) and almost unknown magnetic bacteria.

4.2 Definition of MIC

Microbiologically influenced corrosion (MIC) has been defined in many ways that are more or less similar. Bearing in mind that the term "micro-organism" actually refers to bacteria, cyanobacteria, algae, lichens and fungi [4], some of the definitions for MIC are as follows:

- MIC is an electrochemical process whereby micro-organisms may be able to initiate, facilitate or accelerate corrosion reactions through the interaction of the three components that make up this system: metal, solution and micro-organisms [5].
- MIC refers to the influence of micro-organisms on the kinetics of corrosion processes of metals, caused by micro-organisms adhering to the interfaces (usually called "biofilm"). A prerequisite for MIC is the presence of micro-organisms. If the corrosion is influenced by their activity, further requirements are: (1) an energy source, (2) a carbon source, (3) an electron donator, (4) an electron acceptor and (5) water [6, 7].

MIC is the term used for the phenomenon in which corrosion is initiated and/or accelerated by the activities of micro-organisms [8].

What can be inferred from the above-mentioned sample definitions are the following:

1. MIC is an electrochemical process.
2. Micro-organisms are capable of affecting the extent, severity, and course of corrosion.
3. In addition to the presence of micro-organisms, an energy source, a carbon source, an electron donator, an electron acceptor, and water must be also present to initiate MIC.

We will limit our study in this book to the effect that certain bacteria can have on corrosion. So, in this sense, MIC can be taken as an example of micro-fouling to differentiate it from macro-fouling.[4] However, for reasons that will be described toward

[4] For more on macro-fouling and its effects on corrosion see, for example, [9] and [1]; also especially [10]. In their paper, Palraj and Venkatacahri rank Mandapam first in corrosivity (0.244 mmpy) and third in biofouling. They are also reporting that in their study mild steels exposed to natural seawater for quarterly, semi-annual and annual periods have undergone uniform corrosion.

the end of this chapter, we will define MIC as "*an electrochemical type of corrosion in which certain micro-organisms have a role, either enhancing or inhibiting*".

4.3 Importance of MIC

MIC can occur in almost all environments, such as soil, fresh water, and seawater and all industries such as oil, power generation, and marine industries [11]. MIC is believed to account for 20% of the damage caused by corrosion [12]. On the basis of Gross National Product (GNP), annual MIC-related industrial loss in Australia, for instance, is estimated to be AUD$ 6b [13] (about US$ 5b). A 1954 estimate of MIC loss in buried pipelines, for instance, puts a figure between 0.5 and 2.0 billion US dollars a year, a figure that can only have increased since then [14]. It has been reported that overall loss to the oil and gas industry could be over US$ 100 million per annum [15].[5]

Biocorrosion has been estimated to be responsible of 10% of corrosion cases in the UK [17]. MIC has caused a lifetime reduction of flow lines in Western Australia from the designed 20+ years to less than 3 years [18]. In addition, microbial corrosion has been addressed as one of the major causes of corrosion problems of underground pipelines [19].

Sulphate-reducing bacteria (SRB), a notorious corrosion-enhancing bacteria, has been reported to be responsible for extensive corrosion of drilling and pumping machinery and storage tanks [14, 21]. SRB have also been reported to contaminate crude oil, resulting in increased sulfur levels of fuels. These bacteria are important in secondary oil recovery processes, where bacterial growth in injection waters can plug machinery used in these processes. It has also been suggested that these micro-organisms may play a role in the biogenesis of oil hydrocarbons [14].

MIC failures could have ecological impacts as well such as loss of tritiated D_2O (Deuterium Oxide or Heavy Water) to the environment [22]. Sulphate-reducing bacteria have been responsible for massive fish kills, killing of sewer workers by development of "poisonous dawn fogs", and killing of rice crops in paddies via oxygen depleting [14].

Another interesting application of MIC is in the military, where genetically engineered corrosion-enhancing bacteria could be used to corrode the opposite forces' machinery and facilities so that the logistics of the enemy forces would be paralysed. This aspect, known as "anti-material weaponry", has been discussed in length elsewhere [23].

[5] Tributsch *et al.* quote a work by W.K. Choi and A.E. Torma where in the U.S. industry, an annual loss of about US$200 billion is attributed to MIC, see [16].

4.4 Historical Profile of Advances in Understanding MIC

The role of micro-organisms in corrosion was not investigated until the late 19th century. In fact, several reports of corrosion resembling MIC have been found that date back to the mid-1800s [24].[6] We refer to this era as "historical", (Figure 4.1). During the contemporary era (from the 1920s to the 1960s), MIC was identified and studied. In 1910, Gains considered MIC as an explanation for the very high sulphur content of corrosion products from the Castgill aqueduct in the U.S. In fact, the role of SRB in MIC had been identified even in those early years [25].

More detailed investigations on MIC started as early as 1923 with Stumper's report, to be followed in about 1940 by Starkey and Wight, who indicated that oxidation-reduction (redox) potential was the most reliable indicator of MIC [26].[7] About three years after the discovery of the enzyme hydrogenase[8] in 1931 [25], the first MIC case of failure of underground pipelines was identified [19]. Also in 1934, the first electrochemical interpretation of MIC, proposed by Von Wolzogen Kuhr and Van der Vlugt, provided significant evidence that anaerobic corrosion was caused by the activity of SRB. These two scientists suggested a theory that was named "cathodic depolarisation theory (CDT)", also known as the "classical theory" [28][9].

The years following the introduction of the CDT theory were spent on challenging the theory. As Videla [29] put it, "during the 1960s and the beginning of the 1970s, the research on MIC was devoted either to objecting or to validating" corrosion by SRB as formulated by CDT. It was during these years that electrochemical techniques such as polarisation measurements were applied for the first time in MIC-related studies. While Booth and Tiller produced evidence for CDT [19] in the early 1960s, in 1971 King and Miller minimised the role of SRB in corrosion by putting more emphasis on the corrosion product iron sulphide [28][10]. The mid-1970s produced work by Costello, who introduced an alternative reaction of reduction of biogenic hydrogen sulphide [20]. Costello basically kept Miller and King's theory but instead of hydrogen evolution as the cathodic reaction, he involved hydrogen sulphide produced by the bacteria [29, 28].

Pre-modern times, *i.e.*, the 1980s, may be considered a real "boom" in MIC studies. By the 1980s the impact of stagnant hydrotest conditions on inducing MIC (or more accurately, microbially assisted chloride pitting corrosion) into stainless steel at chloride ion concentrations as low as 200 mg per litre was quite

[6] In this paper, it is reported that in 1891 the role of acids of microbial origin on the corrosion of lead-sheathed cable had been suggested.

[7] It is also interesting to note that Hadley in early 1940s' and Wanklyn and Spruit in early 1950s' were among the first who used open circuit potentials as a function of time for the steel specimens put inside a culture of SRB, see [27].

[8] Hydrogenase is an enzyme that catalyses the reversible oxidation of molecular hydrogen and it is present in many anerobes but it is particularly active in some SRB.

[9] Also see [11].

[10] Also see [11].

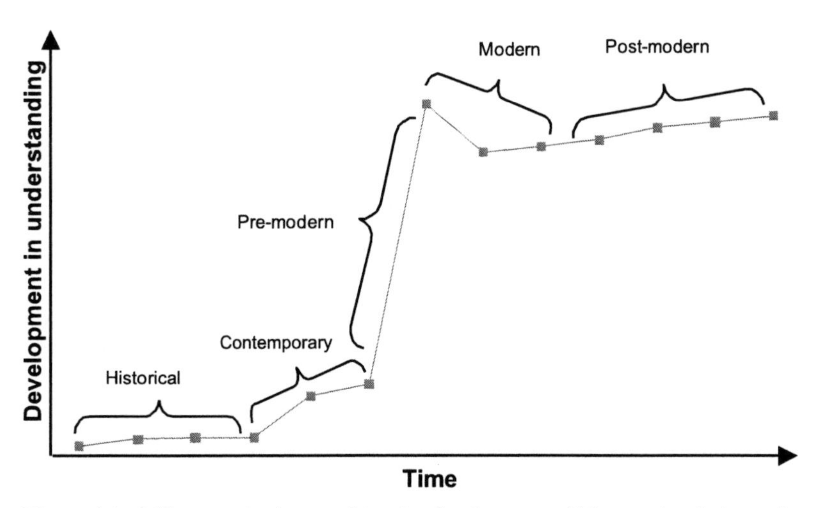

Figure 4.1 Milestones in time marking the development of ideas and techniques for studying MIC

well known [28]. The 1980s also produced the opportunity for more effective communication among almost all disciplines involved in MIC studies, ranging from metallurgy and materials science to microbiology and chemistry. This was enhanced by an increase in the number and quality of experimental studies carried out on MIC. Videla has done a valuable review on this matter [29].

The post-modern era covers the 1990s and beyond. Some of the characteristic activities of this era are the application of rather sophisticated devices such as atomic force microscopy (AFM) in addition to scanning electron microscopy (SEM) and techniques such as energy dispersive X-ray analysis (EDXA) and X-ray diffraction (XRD) [28][11] and electron microprobe analysis in MIC investigations and studies.

In principle, the post-modern era can be said to have the following characteristics [29]:

- Development of new methods for laboratory and field assessment of MIC
- Use of micro-sensors for chemical analysis within biofilm
- Application of fibre-optic microprobes for finding the location of the biofilm/bulk water interface
- Use of scanning vibrating microscopy (SVM) for mapping of electric fields
- Application of advanced microbiological techniques such as DNA probes
- Application of environmental scanning electron microscopy (ESEM), confocal laser microscopy (CSL), and AFM such that the biofilm and its interactions can be observed in real time, allowing profiling of the oxygen concentration within biofilms

[11] EDXA technique detects elements, whereas XRD can be used for crystalline compounds.

I would like to also add that in the 1990s (especially the second half of the 90s and the early years of the following decade), researchers seemingly freed themselves from the paradigm of taking SRB as the most important bacteria in MIC, in contrast to a trend that was predominant during the 1980s. In their iconoclast paper in late 1990s [7], Little and Wagner correctly described such beliefs as "myth". Nowadays, a reasonable amount of work has been generated that considers the effects that bacteria other than SRB can have on corrosion. Examples of such bacteria will be discussed in this chapter, with a particular interest in iron-reducing bacteria.

4.5 Categorising Bacteria

Microbiologists use some "features" to differentiate various types of bacteria from each other. Some of these categorising factors are [30]:

Shape and appearance

1. Vibrio: comma-shaped cells
2. Bacillus: rod-shaped cells
3. Coccus: round-shape cells
4. Myces for filamentous fungi-like cells, *etc.*

Temperature

1. Mesophile: the bacteria that grow best at 20–35°C
2. Thermophile: the bacteria that show activity at temperatures above 40°C

Oxygen consumption

1. Strict or obligate anaerobes, which will not function in the presence of oxygen
2. Aerobes which require oxygen in their metabolism
3. Facultative anaerobes which can function in either the absence or presence of oxygen
4. Micro-aerophiles, which use low levels of oxygen
5. Aero-tolerants, which are anaerobes that are not affected by the presence of oxygen. This means that if these anaerobic micro-organisms are exposed to oxygen, their metabolism will not be literally destroyed by oxygen and they can still be functional.

Figure 4.2 schematically presents the oxygen consumption regimes in a test tube.

Sulphate-reducing bacteria are examples of anaerobic bacteria. whereas sulphur oxidising bacteria are examples of aerobic bacteria (Figure 4.3).

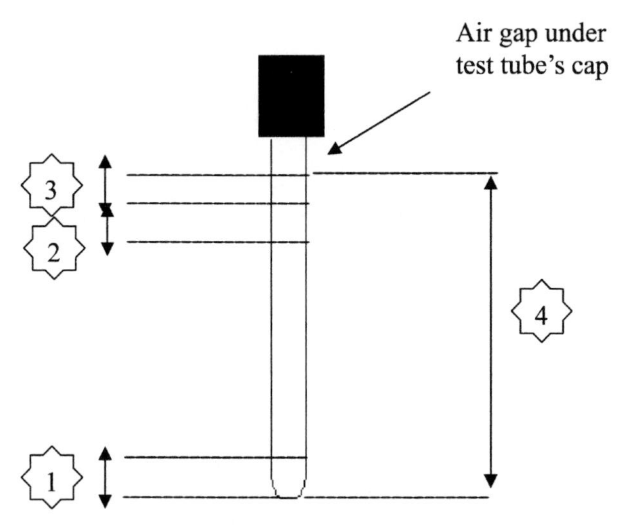

Figure 4.2 Culture development according to oxygen consumption. 1. The zone of strictly anaerobic (Obligate anaerobic). 2. Micro-aerophile band. 3. Aerobic band. 4. The facultative anaerobic zone

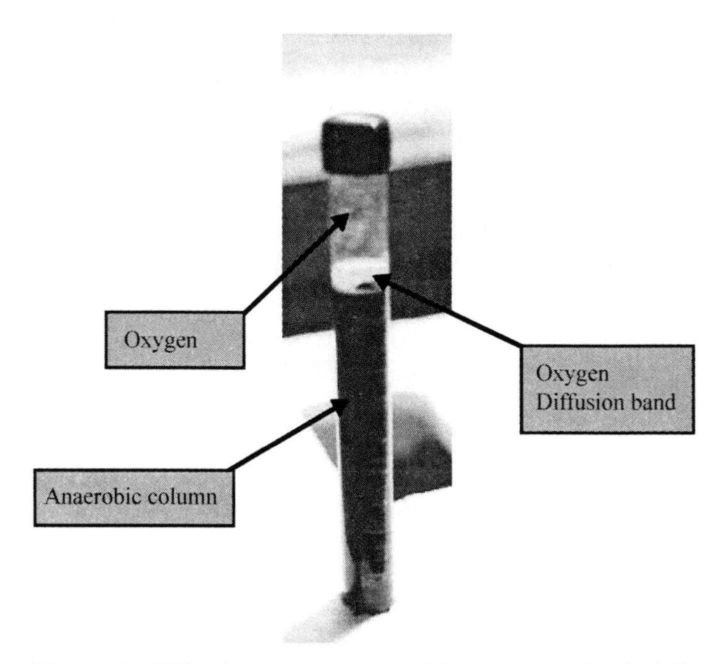

Figure 4.3 SRB culture developed in a solid environment (Agar) within a test tube. A portion of the top section of the sample was taken for transfer purposes. During the culture transfer, oxygen was introduced and diffused into the solid culture. Oxygen did not have a chance to diffuse down further. Note that the bacteria within the oxygen diffusion band are not active, as they are not capable of reducing sulphate and producing the black-coloured iron sulphide

Diversity in Metabolism

1. The compounds from which the bacteria obtain their carbon for growth and reproduction; these can be alternatively called "nutrients"
2. The chemistry by which they obtain energy or recharge the oxidative capacity of the cell; *i.e.*, fermentation or respiration, and the terminal electron acceptors used
3. The compounds they produce as a result of these processes; *e.g.*, organic acids, reduced metal ions, *etc.*

Some facultative anaerobic iron-reducing bacteria can not only reduce ferric ions to ferrous, but can also reduce SO_3^{2-}, $S_2O_3^{2-}$, and S^0 to S^{2-} [31]. Many of the recently described iron reducers are capable of using a variety of electron acceptors including nitrate and oxygen in addition to manganese and ferric ions (Mn^{4+} and Fe^{3+}) [32].

With regard to the energy source, carbon source, and electrochemical reactants, further categorising of the bacterial species is possible. An example of such categorisation [6, 7]can be seen in Table 4.1.

Table 4.1 Categorising bacteria in accordance with the energy and carbon sources and electrochemical reactants

If the ...	... is provided by	... then the growth type is called:
Energy source	Light	Phototrophic
	Chemical Substances	Chemotrophic
Carbon source	CO_2	Autotrophic
	Organic Substances	Heterotrophic
Electron donor (that is oxidised)	Inorganic Substances	Lithotrophic
	Organic Substances	Organotrophic
Electron acceptor (that is reduced)	Oxygen	Aerobic
	NO_2^-, NO_3^-	Anoxic
	SO_4^{2-}, CO_2	Anaerobic

4.6 Biofilm Formation and Its Stages

When bacteria attach themselves onto metallic surfaces, they start to form a thin film known as "biofilm" [30] that consists of cells immobilised at a substratum, frequently embedded in an organic polymer matrix of microbial origin [33]. Biofilms are believed to typically contain about 95% water [34]. Figure 4.4 shows the steps of biofilm formation.

Gradual formation of biofilms can change chemical concentrations at the surface of the metal substrate significantly because the physical presence of biofilm exerts a passive effect in the form of restriction on oxygen and nutrient diffusion to the metal surface.

While a biofilm with a thickness of 100 μm may prevent the diffusion of nutrients to the base of a biofilm, a thickness of just 12 μm can make a local spot anaerobic enough for SRB activity in an aerobic system [35]. Active metabolism of the micro-organisms, on the other hand, consumes oxygen and produces metabolites. The net result of biofilm formation is that it usually creates concentration gradients of chemical species across the thickness of the biofilm [36].

Biofilm formation may take minutes to hours – according to the aqueous environment where the metal is immersed [29]. The first stage of biofilm formation, *i.e.*, the formation of the so-called "conditioning film", is due to electrostatic arrangement of a wide variety of proteins and other organic compounds combined with the water's chemistry, to be followed by the attachment of the bacteria through the EPS to "minimize energy demand from a redundant appendage" [20]. At this stage, the bacteria are referred to as "sessile bacteria" as opposed to their "floating around" or "planktonic" state before attachment to the conditioning film. It has been reported that the presence of sessile SRB on the metal surface results in a higher corrosion rate than that caused by planktonic bacteria alone [37].

When the biofilm is formed and developed, *i.e.*, in stages 1 to 3 in Figure 4.4, the outer cells will start to consume the nutrient available to them more rapidly than the cells located deeper within the biofilm, so that the activity and growth rate of the latter are considerably reduced [37]. Therefore, while the outer cells increase in number, the biofilm starts to act like a "net" to trap more and more particles, organic or inorganic. This will increase the thickness of the biofilm even further.

It is believed that formation of exopolysaccharidic substances (EPS) could help the fragile bacteria as a survival technique to protect themselves from external factors that could be life threatening to them [20] and, perhaps, increasing their capacity to absorb more food by expanding their surface area through the EPS. The role of the EPS material in enhancing corrosion has been emphasised [38].

Under biofilm, factors such as pH, dissolved oxygen, *etc.* may be drastically different from those in the bulk solution, resulting in a phenomenon called ennoblement which has been documented for a range of metals and alloys (*e.g.*, stainless steel at various salinities [30, 33, 39].

Ennoblement can be described as a displacement of the corrosion potential toward more positive potentials [40] that results in increasing susceptibility to pitting, as shown in Figure 4.6. Videla [40] reports that ennoblement involves a change in the cathodic reaction on the metal, caused by the microbial activity within biofilms at the metal/surrounding interface. This phenomenon may serve to clearly justify the effects that biofilm formation can have on changing the electrochemistry of the biofilm-metal system. Controversy still exists regarding the exact mechanism(s) of ennoblement [1], and Dexter has listed the followings as the proposed mechanisms [41].[12]

[12] In addition to these mechanisms, there is a mentioning of "enzymatic mechanism" where hydrogen peroxide (produced as a result of oxidation of glucose) can cause ennoblement of stainless steel, for more details see [42].

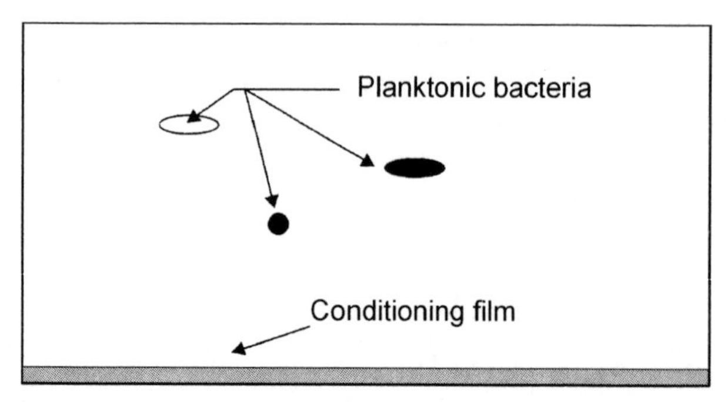

Stage 1: Conditioning film accumulates on submerged surface.

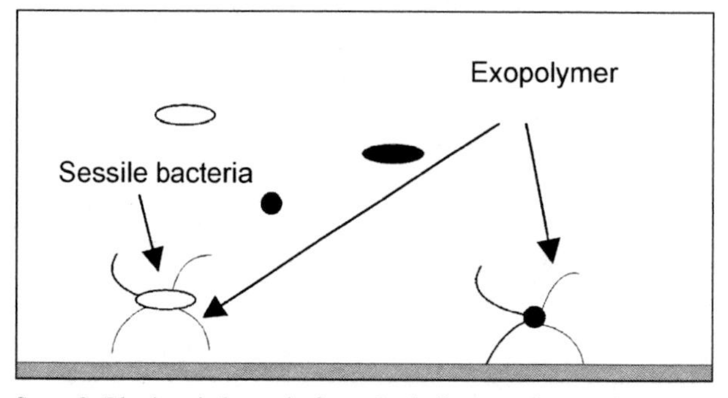

Stage 2: Planktonic bacteria from the bulk water form colonies on the surface and become sessile by excreting exopolysaccharidic substances (EPS) that anchors the cells to the surface.

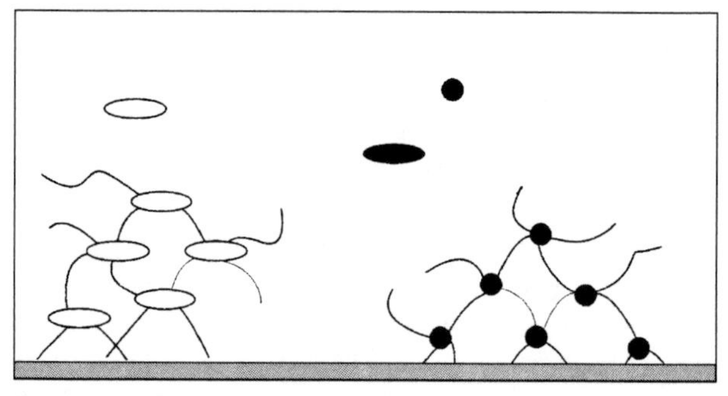

Stage 3: Different species of sessile bacteria replicate on the metal surface.

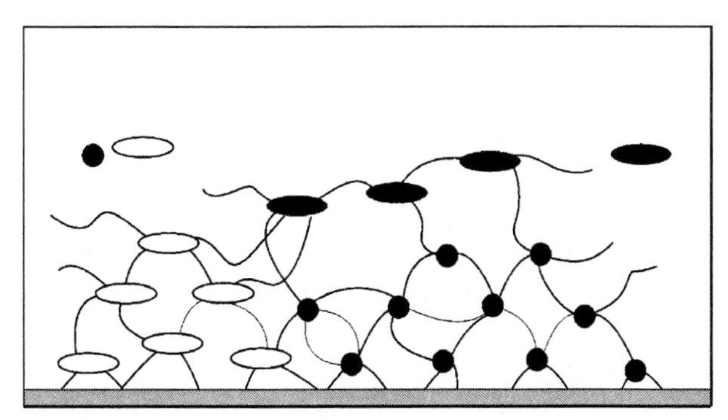

Stage 4: Micro-colonies of different species continue to grow and eventually establish close relationship with each other on the surface. The biofilm increases in thickness and the electrochemical conditions beneath the biofilm begin to vary in comparison with the bulk of the environment.

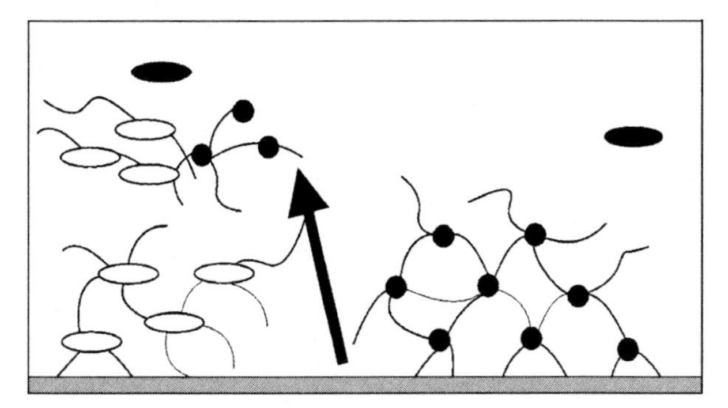

Stage 5: Portions of the biofilm slough away from the surface.

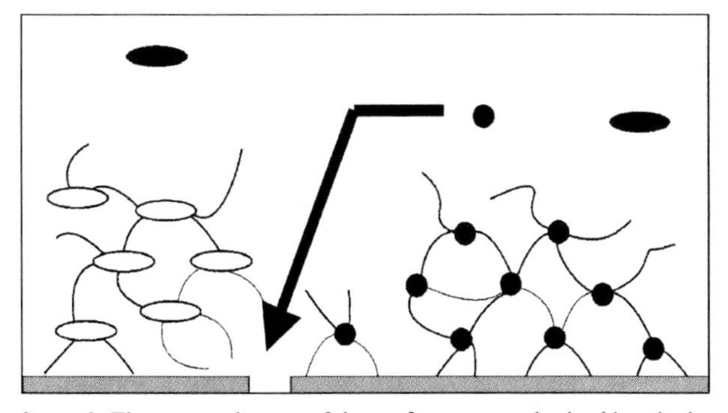

Stage 6: The exposed areas of the surface are recolonised by planktonic bacteria or sessile bacteria adjacent to the exposed areas.

Figure 4.4 Stages of biofilm development [30]

1. Effect of low pH
2. Combination of pH with peroxide and low oxygen
3. Influence of heavy metals
4. Formation of (passivating) siderophores
5. Manganese dioxide contribution

Little *et al.* [1] have pointed out that ennoblement in fresh and brackish water is related to the microbial deposition of manganese whereas in seawater, this phenomenon may be ascribed to depolarisation of the oxygen reduction reaction that may occur, in effect, due to some of the proposed mechanisms mentioned above such as mechanisms 1, 2, and 4. For example, it is well known that the oxygen reduction potential shifts positive (about 60 mV) for each decrease in pH unit and such a decrease produces a noble shift of 35 to 40 mV on stainless steel electrodes in seawater [41, 43], see also [6, 7]. Figure 4.5 shows how the increase in potential due to biofilm formation can endanger the material to pitting.

Corrosion resistance of stainless steels results from formation of a passive oxide film which is stable in an oxidising environment. Any physico-chemical instability of this oxide film either as a result of change in the chemistry of the environment or formation of cracks and/or scratches on the metal surface provides conditions for formation of an oxygen concentration cell which can result in localised corrosion. An example of chemical change of the environment leading into oxide film instability mentioned above is the effect of chloride ions. Chloride ions can locally damage the protective film on stainless steels [44].

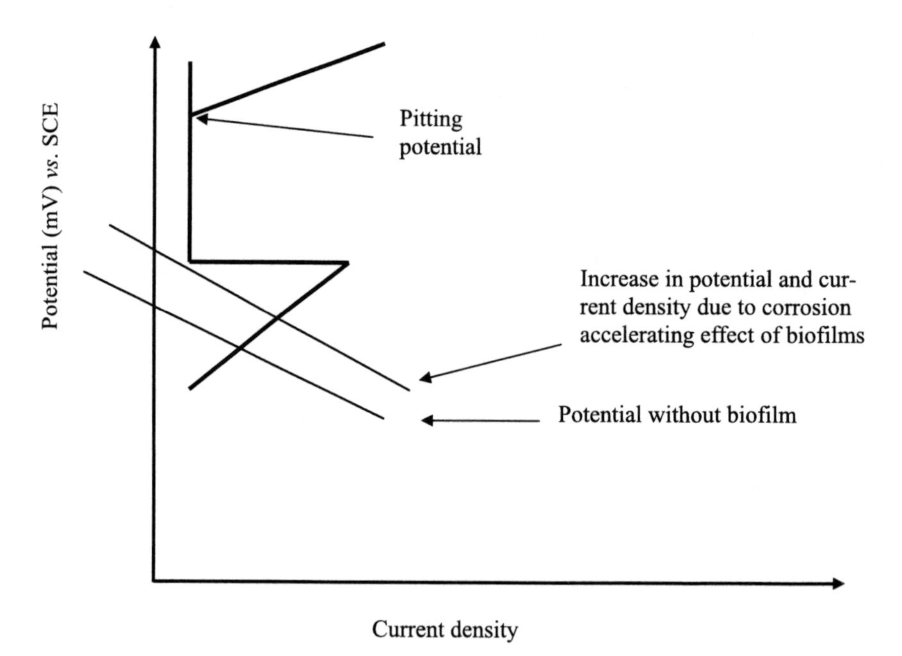

Figure 4.5 Schematic of the effect of biofilm on the ennoblement of carbon steel in the presence of a microbial culture containing corrosion-enhancing bacteria

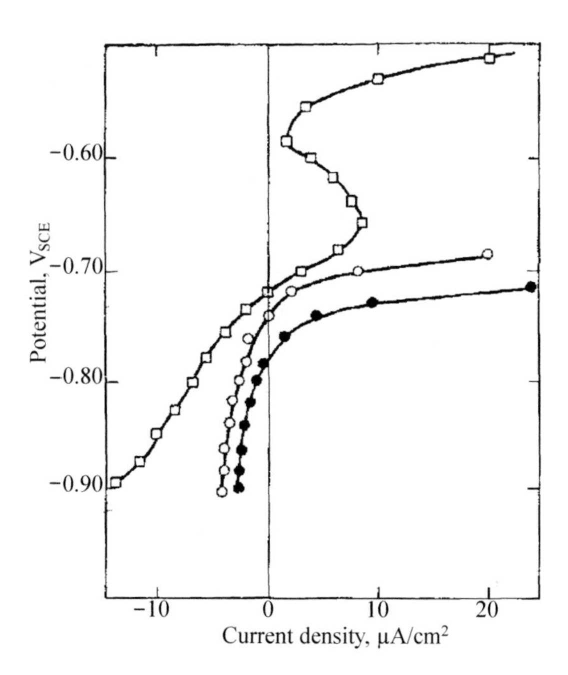

Figure 4.6 How ennoblement increases susceptibility to pitting. Potentiostatic polarisation curves for AISI 1020 steel in anaerobic artificial seawater (pH = 8.0) (□), in artificial seawater contaminated by SRB (total sulfide 10^{-3} M, pH = 7.8, redox potential −510 mV) (○), and in artificial seawater with the addition of 10^{-3} M Na_2S (pH = 8.0) (•).[13] It is seen that the presence of SRB has caused a positive shift (dragging down) the potential, thus facilitating pitting in "lower" potentials

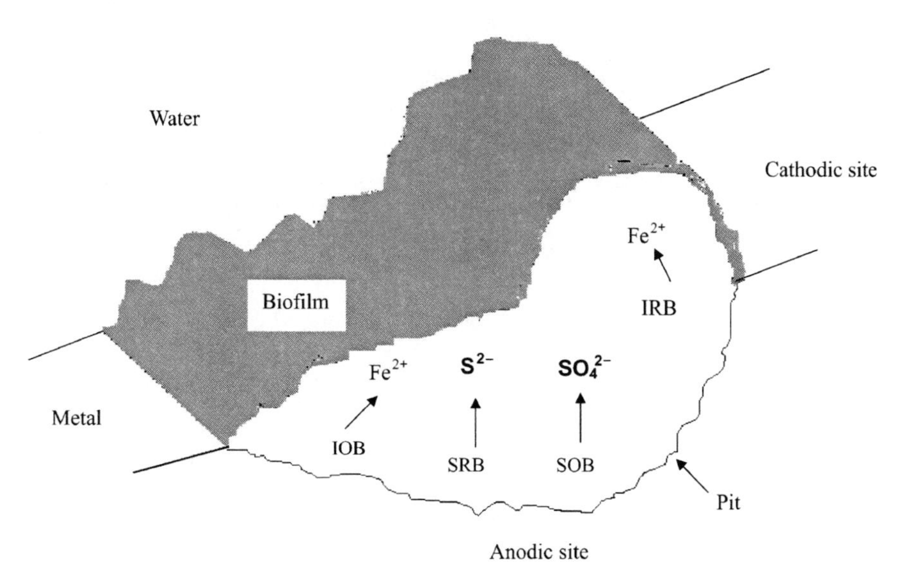

Figure 4.7 Schematic diagram of possible processes that may occur during pitting of steel resulting from biological activity

[13] Reprinted, with permission, from *STP 908 Corrosion Monitoring in Industrial Plants Using Nondestructive Testing and Electrochemical Methods*, copyright ASTM International, 100 Barr Harbour Drive, West Conshohocken, PA 19428. Also see [45].

Steel surfaces can develop biofilms that may form chemical concentration or differential aeration cells resulting in localised corrosion. In addition, if chloride ions are present, the pH of the electrolyte under tubercles (discrete hemispherical mounds [30]) may further decrease, enhancing localised corrosion. In the presence of certain bacteria such as iron-oxidising bacteria (IOB) [32], the chemical conditions under the tubercles formed by the bacteria may become very acidic as Cl^- ions combine with the ferric ions produced by IOB to form a very corrosive acidic ferric chloride solution inside the tubercle [30].

In summary, the bacteria will initiate localised corrosion cells on the inside surface of the tubercles, and the corrosion will progress as a result of the concentration of chlorides induced by bacteria and the low pH generated at the base of the pits [46; 47; 48]. Figure 4.7 shows schematically how bacterial action can induce anodic and cathodic sites leading into pitting. It must be noted that although different types of bacteria are shown in this figure, and in nature it is possible to have different types of micro-organisms living together, it may not be possible for all the bacterial species shown in the figure to co-exist simultaneously.

4.7 How Biofilms Demonstrate Their Effects on Corrosion

Biofilms are contributing to corrosion not only by enhancing the electrochemical conditions and increasing corrosion, but also sometimes by slowing it down. This dual role of biofilms can be puzzling, as it is expected that when bacteria are present in a system, they will form biofilms under which the pits thus produced can be contributing to initiation and/or enhancing of different types of corrosion – for example, stress corrosion cracking (SCC), where local stresses could be built up well above the material's yield point at pits acting as stress concentration sites.

4.8 Enhancing Corrosion

To understand how biofilms can accelerate or decelerate corrosion, an understanding of the structure of biofilms is necessary. Several models have been proposes to explain biofilm structures. Some models are described very briefly below.

4.8.1 *Biofilm Models*

Although MIC and biofilms have been studied for many years, neither the exact mechanisms nor the structure of biofilms are fully understood. Figure 4.8 compares two conceptual models of sulphate reduction for SRB.

According to the classic model of biofilm, due to depolarisation that occurs as a result of sulphate reduction, the anodic reaction becomes more activated, and it net

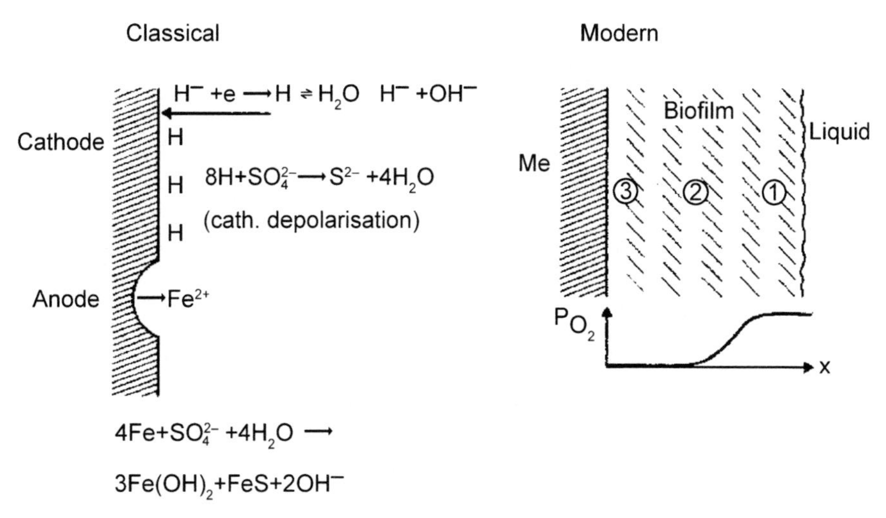

Figure 4.8 Comparison of classic and modern models of biofilm to explain sulphate-reduction [50]

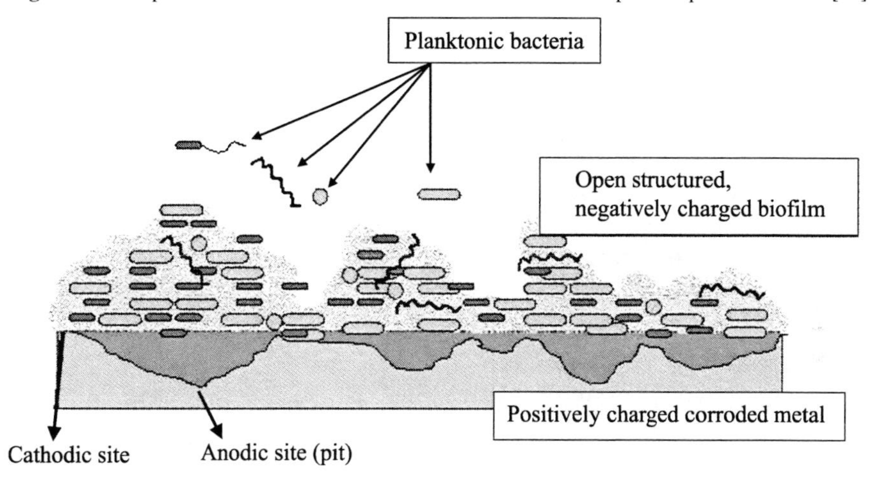

Figure 4.9 A conceptual model for an open, patchy biofilm structure and its regions

result is the production of "rust" in the form of iron sulphide and the creation of an anodic site on the metal substrate. However, new theories have recognised that due to the biofilm build-up, regions near the metal (region 3 in Figure 4.8) are formed that in comparison with regions 2 and 1 are more anaerobic. This may give a good chance for the establishment of an oxygen gradient from the outside of the biofilm thickness toward the inside [49]. Figure 4.9 presents a conceptual biofilm model.

As the model presented in Figure 4.9 shows, the biofilm is a negatively charged, open structure under which localised corrosion can happen. Models describing structure and functions of biofilms have been continuously improving. Some researchers [39, 51, 52, 53, 54] even believe that cell-free biofilms with exopoly-

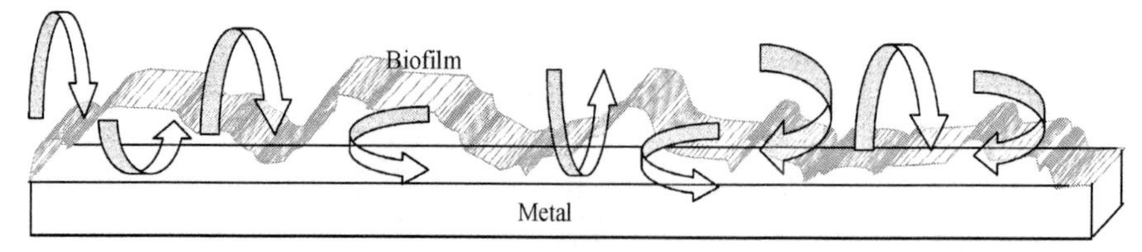

Figure 4.10 An impression of the latest conceptual model of biofilms formed in biotic environments. The arrows mark the entrance and exits of gases (such as oxygen) and chemical species through the "open" structure of the biofilm

mers and function groups, formed within the biofilm, create an environment whose local pH is low enough to favour corrosion.

A more recent model of biofilm assumes a completely open, non-uniform structure where due to non-uniform structure, the establishment of gradients is highly possible [55]. Figure 4.10 presents a cross-section of one such new model.

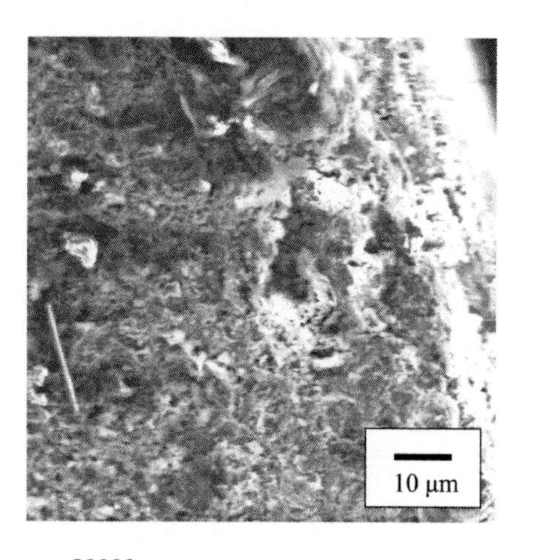

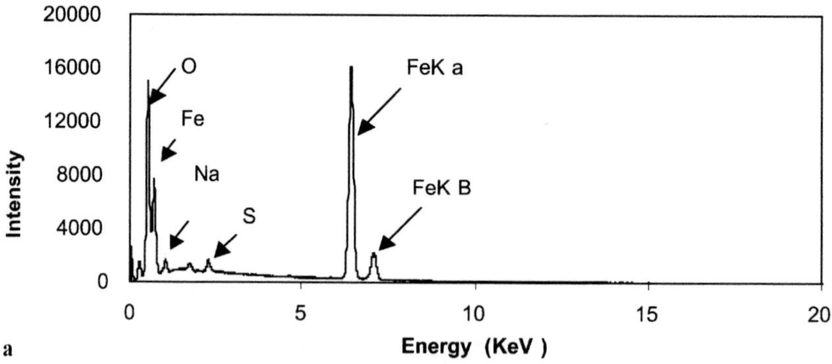

a

Figure 4.11 Comparison of biofilms formed by SRB and IRB. **a** A biofilm formed by SRB (sulphate-reducing bacteria) on carbon steel along with the results of an EDXA analysis of the elements found in it.

The model in Figure 4.10 shows biofilms as an open system where the transport of gases and particles (including chemical species) into and out of the system is quite possible. In such structures, the easy flow of matter and gas transport across the biofilm allows for the establishment of "spots" with high and low concentrations of these chemicals or gases.

When these spots have been formed, differential aeration cells and/or differential concentration cells may be formed. The net results of the formation of such cells are anodic and cathodic sites where anodic sites will manifest themselves as pits. Although this model also allows for transport of gases and materials like in the model presented in Figure 4.8, it emphasises the biofilm as a quite open system rather than layers being laid upon each other with different and distinguishable characteristics. Figures 4.11a and b show examples of biofilms formed by sulphate-reducing bacteria and iron-reducing bacteria on carbon steel. They also

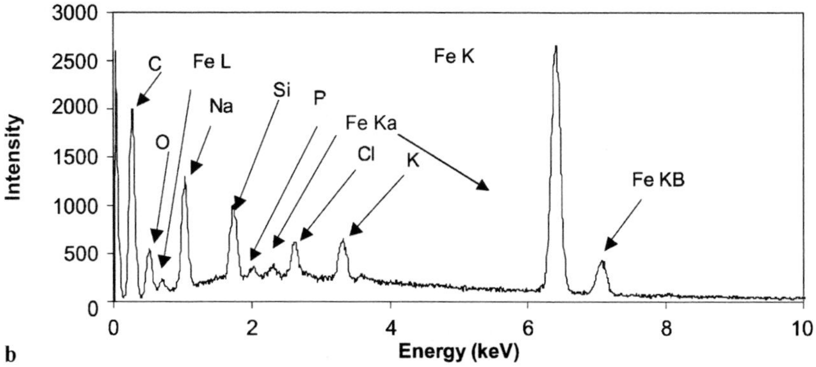

Figure 4.11 Comparison of biofilms formed by SRB and IRB. **b** A biofilm formed by IRB (iron-reducing bacteria) on carbon steel along with the results of an EDXA analysis of the elements found in it (from Javaherdashti R, Making sense out of chaos: General patterns of MIC of carbon steel and bio-degradation of concrete. Proceedings of Corrosion and Prevention 2006 (CAP06), 19–22 November 2006, Hobart, Australia

compare the abundance of elements that have been traced within these biofilms, probably giving rise to the formation and establishment of electrochemical cells such as concentration cells. The patchy fabric of biofilms may result in the formation of differential aeration cells.

4.9 Corrosion Deceleration Effect of Biofilms

Micro-organisms may not always enhance corrosion. The same bacterial species may show both corrosive and protective effects. For example, Hernandez *et al.* [56] reported the corrosive effects of two microbial species, one of which was *Pseudomonas sp.* By changing certain conditions, the very same micro-organisms were showing protective effects and slowing down corrosion. Those researchers also reported that in the presence of bacteria such as aerobic *pseudomonades sp.* and facultative anaerobic *serratia marcescens* in synthetic seawater, corrosion of mild steel is inhibited. The effect seemed to disappear with time in natural seawater. Jack *et al.* [57] reported monocultures of an aerobic *Bacillus* sp. that induced greater corrosion than that of an abiotic environment, but the rate of this corrosion decreased to that of a sterile control after 17 days.

Iron-reducing bacteria (IRB) are a good example of the bacteria that can both accelerate and retard corrosion. These bacteria act by reduction of the generally insoluble Fe^{3+} compounds to the soluble Fe^{2+}, exposing the metal beneath a ferric oxide protective layer to the corrosive environment [58, 59]. *Pseudomonas spp.* are IRB species reported to have corrosive effects [60, 61]. However, there is an increasing body of evidence that IRB could actually slow down corrosion.

Experimental work by Ornek *et al.* [62] has also shown that with biofilm-producing bacteria which can also produce corrosion inhibitors, pitting corrosion of some aluminium alloys could be controlled. It was indicated [63] that two strains of IRB called *Shewanella algae* and *Shewanella ana* were able to significantly reduce corrosion of mild steel and brass. That work postulates that the bacterial strains are capable of reducing the rate of both the oxygen reduction and anodic reactions. Recent research on MIC of mild steel by iron-reducing bacteria [64] has also suggested that this type of bacteria may decrease rather than accelerate corrosion of steel due to reduction of ferric ions to ferrous ions and increased consumption of oxygen. The ferrous ions produced by the bacteria prevent oxygen from attacking the steel surface.

Although Obuekwe had demonstrated the corrosivity of IRB, mainly on mild steel [31, 58, 65], other researchers [39, 51] found that some strains of pure IRB such as *Shewanella* could actually slow down the corrosion process.

The effect of certain conditions has been proposed by some researchers [63, 66]. These "conditions" are schematically shown in Figure 4.12.

The core idea [63] here is that pure IRB can contribute to decelerating corrosion as the ferrous ions produced by the bacteria form a "reducing shield" that blocks oxygen from attacking the steel surface and acts like a protective coating. It

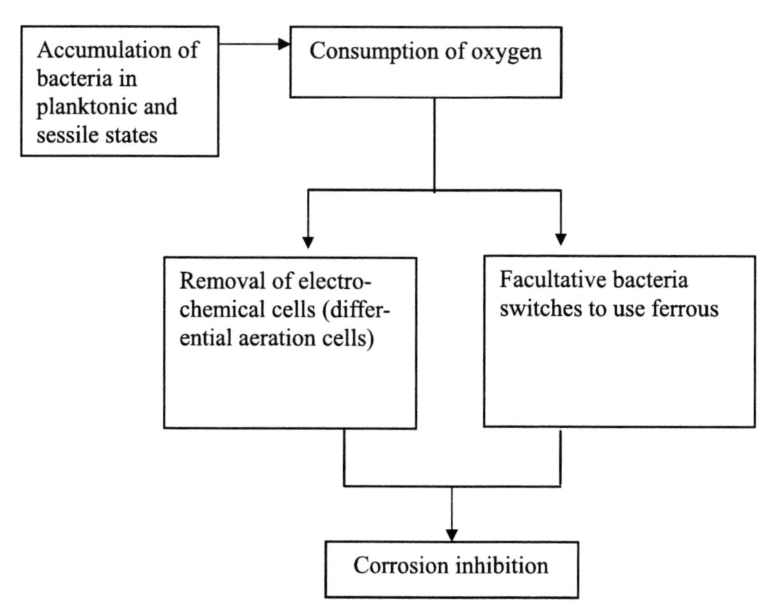

Figure 4.12 The mechanisms occurring in batch systems to inhibit corrosion

seems that this mechanism can happen if the bacterial growth results in biofilm formation on the metal surface. As oxygen is eliminated, for instance by combining with more ferrous ions produced by the bacteria, differential aeration cells are removed. Lee and Newman [66] also suggest that the facultative IRB switch to using ferric iron as the primary electron acceptor. In their view, this in turn will lead to an accumulation of ferrous ions in solution that creates a reducing environment and rapidly scavenges residual oxygen.

Videla [67, see also [40], pp. 74–120, 193–196] has extensively reviewed probable mechanisms by which corrosion can be slowed down or inhibited by bacteria. In this respect, he addresses three main mechanisms that can be summarised as the following:

1. neutralising the action of corrosive substances present in the environment
2. forming protective films or stabilising a pre-existing protective film on a metal
3. inducing a decrease in the medium corrosiveness

Therefore corrosion deceleration could be the result of either one or a combination of these mechanisms. These three mechanisms can successfully explain most of the cases mentioned here. Considering the possibility of having one or more of these mechanisms in place, it seems the bacteria can play a different role in corrosion.

Works by researchers on the slowing down of corrosion by IRB cultures [39, 68] postulate that for batch culture of IRB there is a chance for corrosion deceleration instead of acceleration, due to an increased number of ferrous ions thus produced because of the reduction of ferric ions by these bacteria. These ferrous ions can also combine with oxygen to form more ferric ions and meanwhile deplete oxygen. This can assist in abolishing differential aeration cells and thus decreasing corrosion.

4.10 The Bacteria Involved in MIC

One of the "myths" of MIC, as Little and Wagner call it [7], is the importance of sulphate-reducing bacteria. This is indeed a misleading issue to reduce all MIC problems to SRB by saying "in oil and gas production, the primary source of problems is *Desulfovibrio desulfuricans*, commonly known as SRB" [69].[14] Quoting Sanches del Junco *et al.* [71], it seems that the source of this "SRB myth" started with W.A. Hamilton's work [72] addressing MIC being "most commonly associated with sulphate-reducing bacteria". Certainly, SRB's role has been exaggerated.

Chamritski *et al.* [73] found that MIC of stainless steel 304 in low-chloride (less than 100 ppm) waters could be caused by bacteria such as iron-oxidising

a

Figure 4.13 **a** The biomass formed on a steel pile being exposed to seawater at a depth of 3 m. Note the thickness around the sampling area (courtesy of Extrin Consultants)

[14] It must be noted that the term SRB can not exclusively be applied to address D. desulfuricans only, there are other types of SRB as well. However, *Desulfovibrio* is the most important genera of SRB in salt solutions above 2% (quoted from [70]).

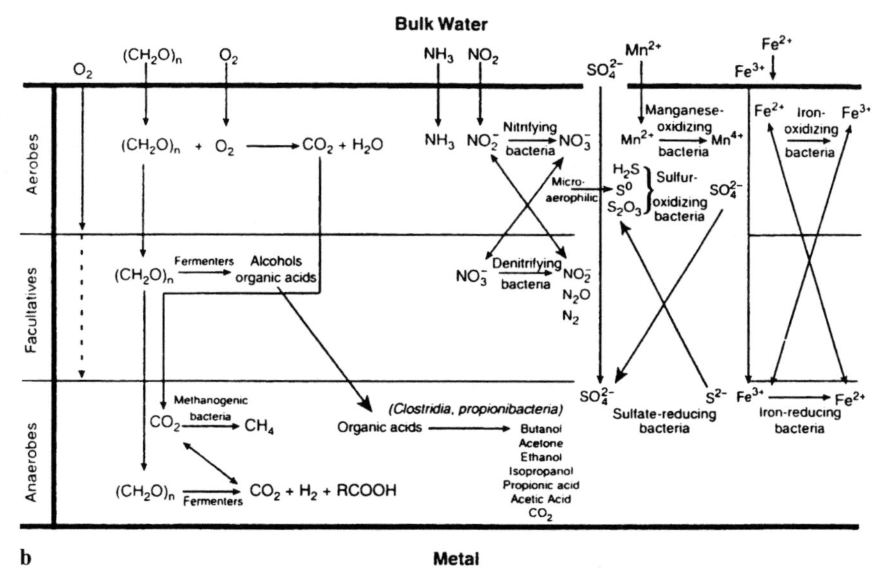

Figure 4.13 **b** Complex environment of a typical aquatic biofilm [7]

bacteria (reduction of the pitting potential), manganese-oxidising bacteria (ennoblement impact), and sulphate-reducing bacteria (pit stablisation effects).

Critchley and Javaherdashti [74], Beech *et al.* [6] and, more completely, Jones and Amy [75][15] give a detailed list of the bacteria that could be involved in corrosion where SRB are just one of these bacterial groups.

In fact, in nature there is no such a thing as a pure culture of this or that bacteria [4], and it is quite possible to have a rather complex picture of all possible microbial reactions that may happen simultaneously or in sequence. Figure 4.13a shows a typical biomass formed on a steel pile being exposed to seawater conditions. Such a mass can easily harbour various types of corrosion-related bacteria. Figure 4.13b is a schematic presentation of possible bacterial types and their interactions within a typical biofilm.

In this section, two examples of the wide spectrum of the bacteria involved in biocorrosion are described: the well-known sulphate-reducing bacteria (SRB) and the relatively unknown iron-reducing bacteria (IRB).

4.11 Sulphate-reducing Bacteria (SRB)

Sulphate-reducing bacteria (SRB) derive their energy from organic nutrients. They are anaerobic; in other words, they do not require oxygen for growth and activity,

[15] Also see [76, 77, 78].

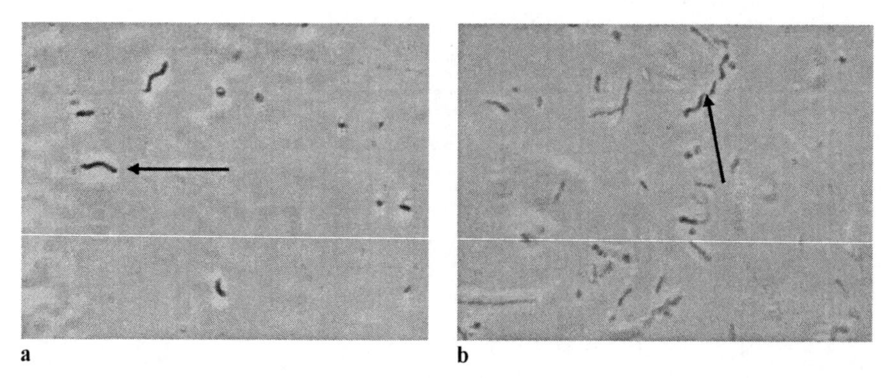

Figure 4.14 Two morphologies of the SRB found in mixed cultures: arrows indicate **a** vibrio **b** spiral

so as an alternative to oxygen, these bacteria use sulphate with the consequent production of sulphide [11].

SRB will grow in the pH range between 4.0 and 9.5 [79]. It has been reported that SRB can tolerate pressures of up to 500 atmospheres [80]. King [20] reported Butlin and Postgate's estimation of sulphide tolerance of SRB to be a concentration of 3000 ppm; however, in another of his work [81], he mentions that the maximum sulphide produced by SRB is not above 600 ppm where the sulphide concentration in sediments and water floods rarely exceeds 500 ppm. SRB can be found everywhere, from more than 70 metres deep in clay [82] to seawater [56]. It is believed that the black colour of the Black Sea could be the result of the activity of these bacteria [83]. SRB can also be found in the human body at sites such as the mouth, [84, 85][16], and bowel [86].[17] By 1997, seven cases of SRB-influenced diseases, two of which were in Australia, had been diagnosed [57] and it seems that this number has increasedg since then[18]. SRB have been reported to be responsible for environmental impacts such as massive fish kills, killing of sewer workers, development of "poisonous dawn fogs" and wastage of rice crops in paddies [14].[19] Figure 4.14a and b show two different morphologies of SRB [89].

[16] Apart from whether or not the SRB are the cause of the mouth malodour, can their existence in the mouth and their known corrosive effects on most engineering materials be a factor in accelerating corrosion of dental fillings?

[17] It has also been reported that 50% of healthy individuals have significant populations of SRB in faeces compared to the 96% of Ulcerative colitis (an acute and chronic inflammatory disease of the large bowel) sufferers especially the *Desulfovibrio* genus, see [87]:

[18] Private communication with Dr. R. McDougal, 18/January/2007.

[19] One must however note that SRB could also have some benefits ranging from assistance in the Evolution [14, pp. 17–19] to contribution to nitrogen-fixing capacity of the soil and killing nematodes which infest the rice plant roots by sulphide toxicity [14, Chapter 8, pp. 205–206].

4.11.1 Mechanisms of MIC by SRB

In 1934 in Holland, VonWolzogen Kuhr and Van der Vlugt provided significant evidence that anaerobic corrosion was caused by the activity of SRB. The two scientists suggested a theory that was named the "cathodic depolarisation theory" or "classical theory". From that time on, modifications to which we collectively refer as "alternative theories" have been made to this original theory.

4.11.1.1 The Classical Theory, Its Rise and Fall

The mechanism postulated by Kuhr and Vlugt attempts to explain the corrosion problem in terms of the involvement of SRB. According to this explanation [28], the bacteria use the cathodic hydrogen through consumption by an enzyme called hydrogenase. It has been postulated that main probable effect of SRB on corroding metal is the removal of hydrogen from the metal surface by means of hydrogenase and catalysing the reversible activation of hydrogen.

Sequences of reactions of the classical theory can be divided into three categories: metal, solution, and micro-organism, as follows:

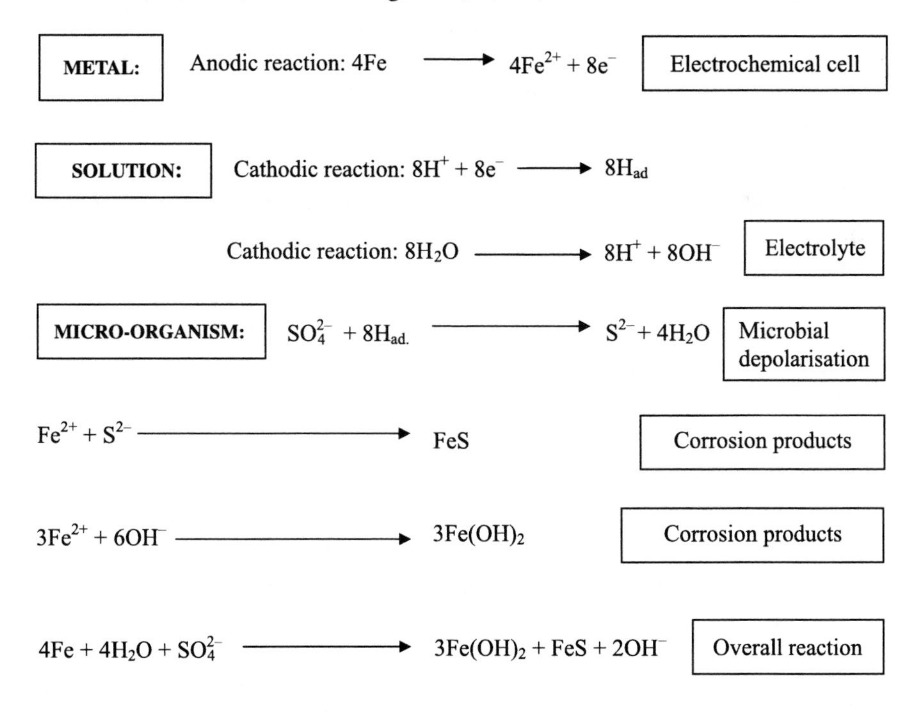

In the absence of oxygen, the cathodic areas of a metal surface quickly become polarised by atomic hydrogen. In anaerobic conditions, the alternative cathodic reaction to hydrogen evolution, such as oxidation by gaseous or dissolved oxygen,

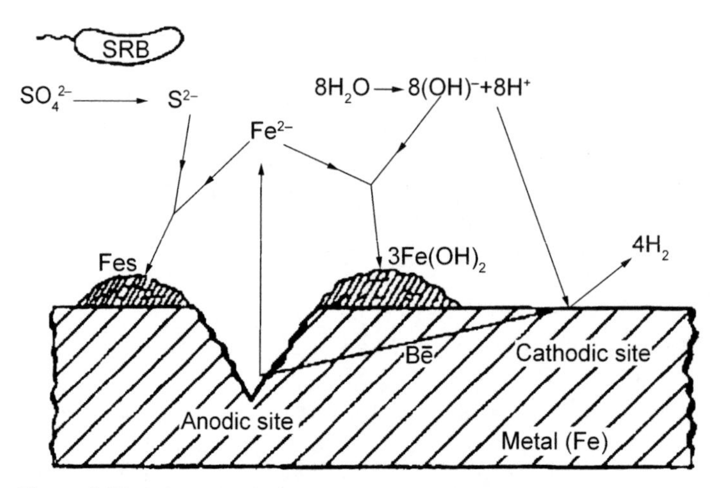

Figure 4.15 Schematic of the cathodic depolarisation "classical" theory of SRB activity [11]

is not available either. These conditions will result in the dissociation of water so as to become the main cathodic reaction with the hydrogen ions thus produced both adsorbed on the metallic surface (polarisation) and consumed by the hydrogenase enzyme. Figure 4.15 schematically summarises the classical theory:

Although the classical theory could explain MIC by SRB for the first time on the basis of electrochemistry, it suffered from serious flaws, some of which are as follows:

1. Research has confirmed that it is impossible for hydrogenase to contribute to the depolarisation of a cathode by removal of atomic hydrogen, as "hydrogenase cannot work on atomic hydrogen at all" [25].
2. According to this theory, the ratio of corroded iron to iron sulphide must be 4:1, however, in practice this ratio varies from 0.9 to 1 [90].
3. A recent study [18] used a culture of nitrate-reducing SRB that could grow and consume hydrogen faster and more efficiently. When sulphate was replaced by nitrate, these nitrate-reducing bacteria proved to efficiently oxidise the cathodic hydrogen from the metal, but unlike sulphate-reducing bacteria cultures, they failed to stimulate corrosion. Thus, these data showed that MIC by SRB could not be attributed merely to the uptake of cathodic hydrogen.

4.11.1.2 Alternatives to the Cathodic Depolarisation Theory

The discoveries of the shortcomings described above helped shift the paradigm of involvement of SRB in the corrosion to that which collectively can be called "alternative theories". These theories cover a wide range of research whose main common point is their attempt to explain MIC by SRB although not directly involving the bacteria itself.

As Stott reports [25], as early as 1923, Stumper had shown that the metal sulphides themselves can act like cathodes to the underlying steel, thus generating a galvanic cell and increasing the corrosion rate, even in the absence of hydrogen sulphide. In 1971, Miller and King attributed the corrosive effect to both hydrogenase and the iron/iron sulphide galvanic cell [25]; in other words, they proposed iron sulphide as the absorber of molecular hydrogen [91]. This was in fact the first step towards minimising the role of the bacteria in cathodic depolarisation [29]. A modification to Miller and King's proposal was made in mid-1970s by Costello, who replaced iron sulphide with hydrogen sulphide as the cathodic reactant as shown in the reaction below:

$$2H_2S + 2e^- \rightarrow 2HS^- + H_2$$

In addition to these theories, Iverson proposed a hypothesis regarding the existence of a corrosive phosphorous metabolite leading to observed high corrosion [92].

New theories put more emphasis on the anodic breakage of iron sulphide films and the galvanic cell formation in anodic spots and zones that have an enhanced SRB population [5]. Videla summarises the new picture of the SRB-induced MIC mechanisms as follows [93]:

- In saline media, at high Fe^{2+} concentrations, the steel is dissolved, resulting in the formation of a hydrated ferrous hydroxide film where the thickness and protective characteristics of this film depend on factors such as the concentration of Fe^{2+} and the solution's acidity (pH)
- The anion adsorption processes that are occurring at the metal/solution interface will be competing with each other, so that the outcome of these competitions could either be enhancing or inhibiting corrosion
- The physico-chemical properties of the iron sulphide film can control the impact of sulphides on the steel dissolution, whereas these impacts and effects themselves are dependent on the ferrous ion/sulphide anions ratio, the presence of SRB and how the biofilm has covered the metal surface.[20]

As seen in all of these new theories, despite their similarities and dissimilarities, the role of the bacteria in corrosion becomes less and less important. Recent research by Hang [95] revealed very interesting results. In this research, SRB were directly enriched with metallic iron and sulphate as the only growth substrate in carbon dioxide/bicarbonate-buffered medium. The rod-shaped SRB isolated from the culture have been shown to be genetically very closely related to *Desulfobacterium catecholicum,* yet physiologically significantly different from them! This

[20] It may be worth noting that researchers such as Smith and Miller in their review of the corrosive effects of sulphides on ferrous metals have reported that in media with high ferrous ion concentration, most of the corrosion of mild steel in biotic (bacterial) cultures can be attributed to the ferrous sulphide produced by the bacteria. In other words, it seems that when SRB are present, the iron sulphide produced by their interactions could be more corrosive than chemically (no bacteria) prepared iron sulphide. See [94] (The author thanks Dr. Peter Farinha's for his remarks regarding this paper and his kindness in providing the author with this paper).

new species has been given the name *Desulfobacterium corrodens.* But this is not the whole story; the bacterial strains use only iron, lactate, and pyruvate for the reduction of sulphate. In the presence of iron, the strain reduces sulphate more rapidly than *Desulfovibrio,* whereas in the presence of hydrogen or lactate, sulphate reduction becomes remarkably slower than for the *Desulfovibrio* species. This work also reports another new species of *Desulfovibrio* (named *Desulfovibrio ferrophilus*) that, in the presence of iron, could reduce sulphate at a higher rate than other *Desulfovibrio* species but slower than *Desulfobacterium corrodens.*

In this study, [95] model anaerobic corrosion of iron without the involvement of hydrogen. They postulate that the SRB that grow in very close contact with the iron surface can take electrons directly from the metal surface (in a stept they call "electron pick-up") and transfer these electrons to the sulphate-reducing system (SRS). While this proposed mechanism is certainly a breakthrough, there are still serious questions to be answered. For example, it is unknown how the electron pick-up step works and what mechanisms are involved there. As we will see later, Little *et al.* [32] have also demonstrated that for another group of bacteria which are important in corrosion (*i.e., Shewanella purefaciens* which are iron-reducing bacteria), the reduction of metal requires contact between the cell and the surface where the reduction rate is directly related to the surface area. The same researchers also found that the location of pits induced by these bacteria on carbon steel coincided with sites of bacterial colonisation.

One cannot help but think that if Hang's approach is correct, then all the alternative theories that so far have tried to minimise the role of SRB in MIC would have to be seriously reconsidered.

4.11.2 Examples of Corrosion by SRB

Almost all types of engineering materials have been reported to experience MIC by SRB; copper, nickel, zinc, aluminium, titanium and their alloys [96, 97, 98], mild steel [72, 99, 100], and stainless steels [68, 80, 101, 28] are just some examples. Among duplex stainless steels, SAF 2205 has been reported for its vulnerability to MIC [44, 102, 103]. According to these studies, SAF 2205 can corrode and have pitting initiated due to the presence of SRB after immersion in seawater for more than one year (18 months) [102]. Corrosion rates of 10 mm/year [5] in oil treatment plants and 0.7 mm/year to 7.4 mm/year due to the action of SRB and/or acid-producing bacteria in soil environments [8] have been reported.

4.11.3 Stress Corrosion Cracking (SCC) and SRB

Stress corrosion cracking (SCC) is a type of corrosion that is caused by simultaneous action and effects of both tensile stresses on a vulnerable material in a corrosive medium.

Gradual formation of biofilms can change chemical concentrations at the surface of metal substrata significantly: The physical presence of a biofilm exerts a passive effect in the form of restriction on oxygen diffusion to the metal surface. Active metabolism of the micro-organisms, on the other hand, consumes oxygen and produces metabolites. The net result of biofilm formation is that it usually creates concentration gradients of chemical species across its thickness, which is typically between 10 μm to ~400 μm [36].

If chlorides are present, the pH of the electrolyte under the biofilm may further decrease, leading to more severe corrosion. When some types of bacteria such as iron-oxidising bacteria (IOB) [32] are present, the tubercule conditions may become very acidic due to combining of the chloride ions with the ferric ions that are produced by the bacteria to form acidic ferric chloride solution inside the tubercule (or biofilm) that is highly corrosive [30]. Pitting is the predominant morphology of MIC [21, 46, 104].

On the other hand, pitting can act as an SCC initiator; because the "roots" of pits act as "stress magnifiers", so that the applied stress becomes multiplied several times, resulting in stresses far in excess of the tensile yield strength, thus producing failure [105].

Among the investigations addressing the effect of SRB and other bacteria such as iron-reducing bacteria (IRB) on enhancing corrosion of steels (carbon steel, stainless steel 316, and duplex stainless steel SAF2205), Javaherdashti *et al.* produced a series of papers [13, 106–109]. In these studies, mixed (containing SRB, IRB, and other unidentified micro-organisms) and pure cultures of SRB (only SRB) and IRB (only IRB) and their impacts on both electrochemical and mechanical properties of the above-mentioned steels were investigated. The test cell used for conducting SCC by slow strain rate testing (SSRT) for the steel samples had been designed in such a way that it could sustain the environment anaerobic enough for the SRB. For this reason, the test chamber was designed such that it could reveal blackening as a sign of growth (Figure 4.16a). The SRB biofilm could easily be observed (Figure 4.16b).

It is interesting to see how mixed and pure cultures of SRB can affect the severity of SCC of carbon steel and duplex stainless steel by decreasing the time of failure. In other words, when SRB is present, the material is likely to fail in a relatively shorter time than an abiotic (no bacteria present) environment (Figures 4.17 and 4.18a,b).

a

b

Figure 4.16 **a** SSRT of a carbon steel sample in the anaerobic chamber inoculated with SRB. Note the oil layer (arrow) to prevent oxygen ingress [106]. **b** Close-up of Figure a showing thick, black biofilm formed on the exposed section of the mild steel SSRT sample [109]

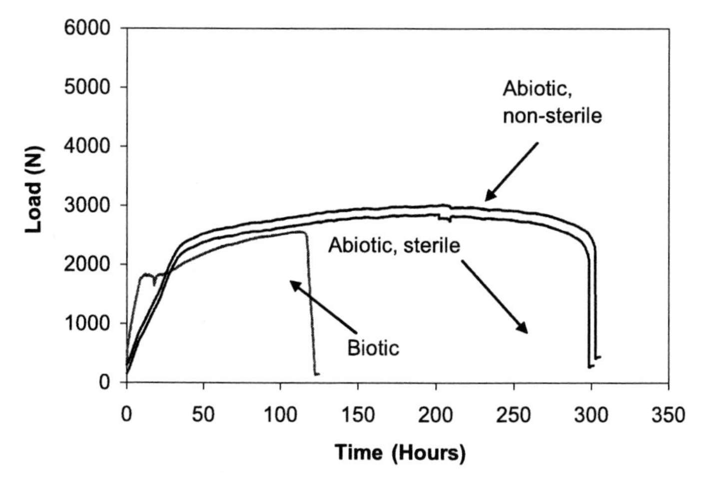

Figure 4.17 Typical load *vs.* time curves generated by SSRT tests of mild steel in the environments consisting of a mixed SRB culture, abiotic non-sterile containing 3.5% sodium chloride solution alone, whereas the abiotic sterile environment contained modified Postgate B medium along with some chemicals to keep it sterile [106]

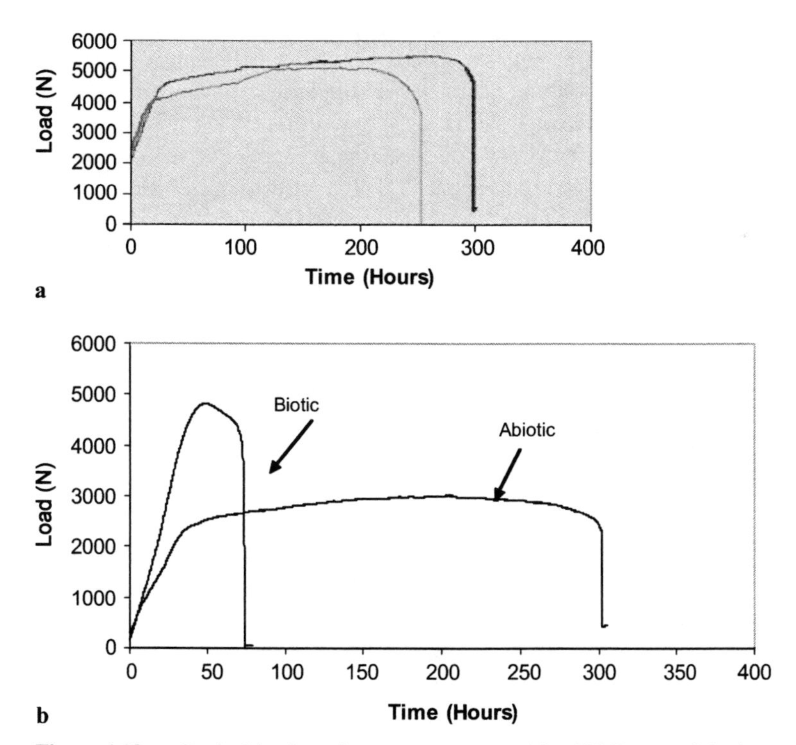

Figure 4.18 **a** Typical load *vs.* time curves generated by SSRT tests of duplex stainless steel SAF2205 in abiotic and biotic (mixed culture of SRB) environment [108]. **b** Typical load *vs.* time curves generated by SSRT tests of mild steel in a 3.5% chloride solution, with and without pure SRB culture, termed respectively biotic and abiotic conditions [106]

4.12 Iron-reducing Bacteria

There are other micro-organisms in addition to SRB which are also important in corrosion. For example, the MIC of stainless steel 304 in low-chloride natural water can involve the combination of some or all of the following factors [73]:

- Ennoblement of potential, possibly caused by manganese-oxidising bacteria
- Reduction of the pitting potential because of (1) the crevice-like action of surface deposits produced by iron-oxidising bacteria, or (2) the activating effect of sulphide or thiosulphate produced by SRB, or (3) simply the effect of silicate in the water.

Iron-reducing bacteria (IRB) are another group of micro-organisms which are of interest in MIC. However, it seems that their importance in corrosion has been over-shadowed by the iron bacteria (IB), or more precisely, iron-oxidising bacteria (IOB). For example, ASTM D 932-85 defines iron bacteria as a general classification for micro-organisms that utilise ferrous iron Fe^{2+} as a source of energy, and are characterised by the deposition of ferric Fe^{3+} hydroxide [110]. A common example of IOB is the *Gallionella* sp.

The reducing effects of IRB on metals such as copper, nickel, gold and silver have been known for nearly 50 years [111]. As the name implies, IRB act by reduction of the generally insoluble Fe^{3+} compounds to the soluble Fe^{2+}, exposing the metal beneath a ferric oxide protective layer to the corrosive environment [31, 58, 64].

It is important to understand how iron-reducing bacteria can reduce iron, or more precisely, ferric iron ion. The reason is that while the bacteria can reduce iron in some way or another, it is one of these methods that may be of more importance with regard to its contribution to corrosion. In the following section, possible reasons and mechanisms for microbial iron reduction are discussed.

4.12.1 Why Is Microbial Reduction of Iron Important?

Some possible reasons why iron reduction by bacteria is important:

1. Availability of iron: Iron is not very soluble, but if it is reduced to ferrous iron (which is soluble) so that the organic compounds can stabilise iron by chelation where, later on, that iron can "liberate" itself from the organic matter and precipitate as iron [32, 112].
2. IRB are a very important part of the soil microbial community, as most of the IRB are facultative anaerobes, and thus if oxygen is available, they will prefer it for their growth while also maintaining their capability of growth under anaerobic conditions too. It is estimated that in the surface layer of soil, on the average, the number of IRB could be as high as 10^6 cells per gram of soil [113]. It must be kept in mind that as IRB are both chemoheterotrophic (organic compounds are the source of energy for them) and facultative anaerobes, their

numbers within the soil's surface layer is higher than at deeper levels, especially if the soil is rich in organic matter at the surface level [112]. As a result, in cases in which their numbers in soil are reported, the depth of sampling for the organic carbon content must also be recorded.

3. Incorporation (assimilation) of iron into proteins containing heme or iron-sulphur [40, 66, 67].
4. IRB are capable of making the environment suitable for SRB. In a mixed population of micro-organisms in a biofilm, as oxygen is consumed, the redox potential starts to decrease so that nitrate, then manganic and ferric ion and the sulphate are reduced [112]; this consequence can be seen in Table 4.2.

Most of the IRB are fermentators under anaerobic conditions, however, there are a few that actually need ferric iron under anaerobic conditions [112]. To add more into the complex picture, some of the IRB can use nitrate for anaerobic respiration [112]. Little *et al.* reported that IRB such as *Shewanella purefaciens* can use oxygen, Fe(III), Mn(IV), NO_3^-, NO_2^-, $S_2O_3^{2-}$, SO_3^{2-} and others [32]. The same researchers also reported that *S. purefaciens* under aerobic and anaerobic conditions may or may not use the same material (*e.g.*, acetate that can be used aerobically but not anaerobically). Perhaps Panter is right in his recommendation that "oxygen content [for IRB] is more important in determination [of their] numbers than available ferric ion content" [112].

In soil environments, most IRB that can be isolated are fermentators, and for the IRB that carry out dissimilatory reduction of ferric ion by anaerobic respiration, isolation may not be "as regular"; however, the latter can more easily be isolated from freshwater streams, lakes and marine waters [112]. Javaherdashti [89] isolated a *Bacillus* sp. that could grow in nutrient broth under aerobic conditions. The bacterium was also motile in Postgate B medium modified with 35 g/l NaCl. This isolate was from a muddy sample taken from the depth of 14 m of the Estuary of Merimbula River, New South Wales, Australia. Figure 4.19 shows such a bacterium.

In fact, the mechanisms of microbial iron reduction can be grouped into two [40, 66, 67]:

- Assimilation
- Dissimilation

Table 4.2 Sequence of reduction in redox potential (E_h) under anaerobic conditions [50]

....	... is reduced to	Comments	E_h
NO_3^-	N_2	Through first reduction of NO_3^- into NO_2^- and then into N_2O	< 400 mV
NO_3^-	NH_4^+	By first reduction of NO_3^- into NO_2^-	
Mn^{4+}	Mn^{2+}		< 400 mV
Fe^{3+}	Fe^{2+}		< 300 mV
SO_4^{2-}	H_2S		< 100 mV
Organic C	H_2, CO_2		< −100 mV
$H_2 + CO_2$	CH_4		< −300 mV

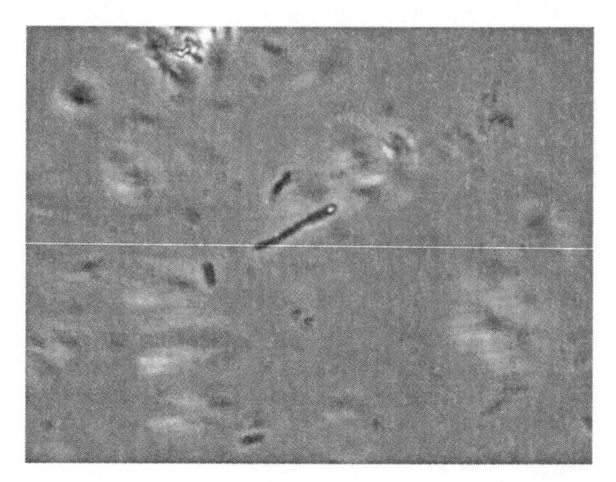

Figure 4.19 Iron-reducing bacterium culture. The terminal bright spot is an endospore (1000X) [73]

Assimilation, as mentioned above, is unlikely to have an effect on corrosion as trace amounts of iron are required for it to occur, whereas dissimilatory iron reduction involves electron transfer to iron as part of both anaerobic fermentation or anaerobic respiration [40, 66, 67]. The impact of fermentor iron reducers has not been studied in detail, perhaps because they do not reduce ferric iron as rapidly or extensively as anaerobic respitory IRB [112]. However, Panter reported that fermentative IRB in submerged environments are encountered more frequently than the IRB that use ferric ion in anaerobic respiration [112]. Nonetheless, as mentioned earlier, it is not yet known whether the fermentative IRB could have a great contribution to corrosion. Most probably, then, the only remaining nominee for having an impact on corrosion would be the respiratory iron reducers.

IRB are very interesting when considered for their effects on corrosion. The next section considers their impact on the corrosion severity.

4.12.2 Contradicting Impacts of IRB on Corrosion

Most engineers and even scientists who are familiar with MIC would not believe that sometimes the bacteria can actually retard corrosion and protect the metal. In fact, there is a growing body of evidence that IRB can, under some circumstances, enhance corrosion, and under other circumstances, inhibit corrosion.

In the following sections, examples of corrosion enhancement by IRB are presented. The next section includes an overview of some possible reasons for the IRB to inhibit corrosion.

4.12.2.1 Corrosion Enhancement by IRB

Obuekwe *et al.* in a series of papers on IRB (*Pseudomonas* sp.) reported corrosion effects of the bacteria under the micro aerobic (which contains trace amounts of oxygen) conditions [31, 58, 64]. These works included polarisation studies of mild steel in media with and without yeast extract. Those researchers reported that the IRB may contribute to corrosion of mild steel by anodic depolarisation due to their ability to reduce and remove the protective film of ferric compounds.

Obuekwe's pioneering work on characterising corrosion effects of IRB by using a polarisation method has been debated, as a potentiodynamic approach over a range of 0.4 V was used to examine corrosivity, and this may affect and alter the "natural" behaviour of microbial communities.

The examples below suggest how "opposite" results may be obtained by applying voltage:

- A report on the CP effects on steel pipes against MIC [114] suggests that under laboratory conditions, applying voltages more negative than $-0.98V_{Cu\text{-}CuSO_4}$ may decrease the number and/or the activity of iron bacteria as a result of environmental changes caused by cathodic protection process. Although in this report, the type of the bacteria (IOB or IRB) has not been specified, from general recognition of iron bacteria [110], it may be anticipated that it was iron-oxidising bacteria whose number had been adversely affected by applying voltage. The report thus demonstrates the negative effect of applying voltage on micro-organisms and their numbers.
- It has been recommended practice to apply a voltage of about $-0.98V_{Cu\text{-}CuSO_4}$ in order to suppress bacterial effects by cathodic protection, resulting in a decreasing extent and severity of corrosion. In this way, the localised pH is increased and the environment becomes too alkaline for the micro-organisms to comfortably withstand, thus decreasing the corrosion rate. However, in one particular case of cathodic protection, it has been reported that applying voltages up to $-1.1V_{Cu\text{-}CuSO_4}$ not only failed to prevent the growth of bacteria on the metal surfaces, it rather prompted the growth of certain microbial species and the rate of corrosion [115]. The possible effects of CP on MIC will be discussed in more detail in Chapter 10 of this book.

The same debatable effects might have also affected the results in the work by Obuekwe. It seems that applying a voltage to the medium (as was done in Obuekwe's works on corrosion of mild steel by IRB) may not resemble MIC properly because there is no way to know how the microbial activity has been affected by the applied voltage and how this would affect the outcome of the experiments.

On the other hand, Little *et al.*, who did not use polarisation methods but instead one of the safest electrochemical methods, electrochemical noise analysis (to be discussed later in Chapter 6) reported the corrosion-enhancing effects of another type of IRB, *Shewanella purefaciens* [32].

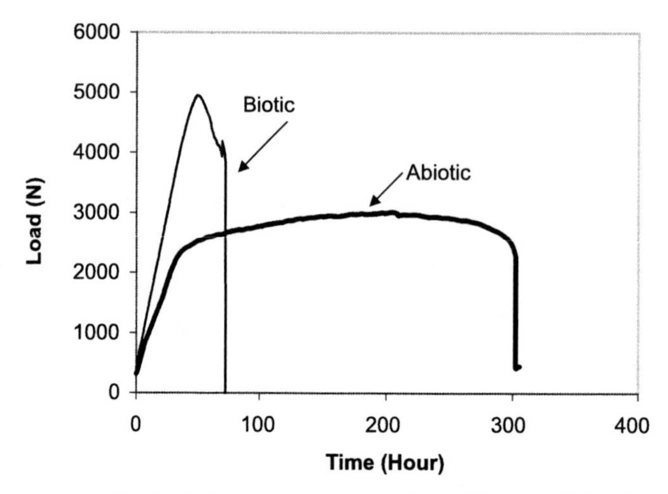

Figure 4.20 Typical load *vs.* time curves for mild steel in IRB culture, comparing it with slow strain rate behaviour of mild steel in abiotic synthetic seawater media

Javaherdashti in his investigation regarding the mechanical and electrochemical behaviour of mild steel, stainless steel 316L, and duplex stainless steel SAF2205 found that when mild steel is exposed to a culture of IRB, in comparison with an abiotic environment it shows lesser times of failure, therefore implying that IRB could actually enhance corrosion [89]. Figure 4.20 illustrates the typical slow strain rate SCC behaviour of mild steel in a culture of IRB.

The above-mentioned points may suggest that IRB are indeed important in increasing the corrosion rate. If you have a mixed culture of SRB and IRB, for example, the carbon steel sample in the mixed culture will fail earlier with respect to an abiotic environment (Figure 4.17). A possible explanation for premature failure of mild steel in such a mixed culture is shown schematically in Figure 4.21.

However, IRB still have the power to surprise us! Lee *et al.* reported that a mixed culture (biofilm) containing IRB (*Shewanella oneidensis*[21]) and SRB (*Desulfovibrio desulfuricans*) that had been formed on mild steel could provide a short-term (four days) protection to the steel [116]. As the authors put it, "[t]he fact that an iron-reducing bacterium can inhibit corrosion when a corrosion-enhancing bacterium is present warrants future study with respect to its potential applicability to the design of biological corrosion-control measures". Such reports can lead us into another aspect of IRB: a corrosion-inhibiting bacteria! This matter is discussed in Section 5.2, "Corrosion deceleration effect of biofilms" of Chapter 5 and will not repeated here.

[21] *Shewanella oneidensis* is a facultative anaerobe that can use oxygen or ferric ion as its terminal electron acceptor. [63].

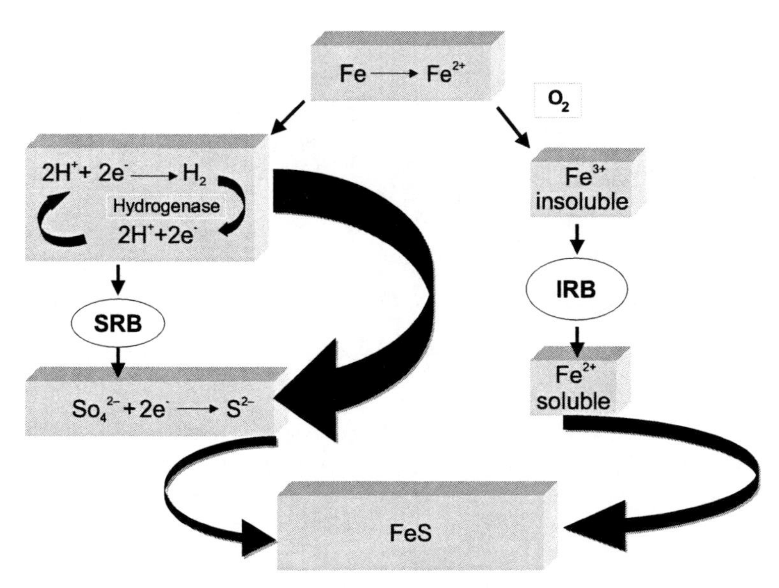

Figure 4.21 Possible interaction between SRB and IRB

4.13 Magnetic Bacteria

Magnetic bacteria have the ability of synthesising intracellular nano-sized fine magnetic particles [117]. Each of these magnetic particles, called a magnetodome, is about 50 nm in width [118]. Figure 4.22 is a schematic presentation of *Aquaspirillum magnetotacticum* where magnetosomes can be clearly seen as a string. Note that the total magnetic energy of the magnetosome string is the sum of the individual magnetic moments of the beads, so magnetic energy of the cell is calculated as being in the order of 10^{-19} J/G, adequate to align the bacterium in the 0.5 G geomagnetic field [118][22].

First discovered in 1975 by Blakemore, the magnetotactic bacteria are bottom-dwelling micro-organisms which are either anaerobic or microaerophilic [88]. It seems that the tendency of the bacteria to migrate downward along the component of the magnetic field is an evolutionary tactic that the anaerobic bacteria use to

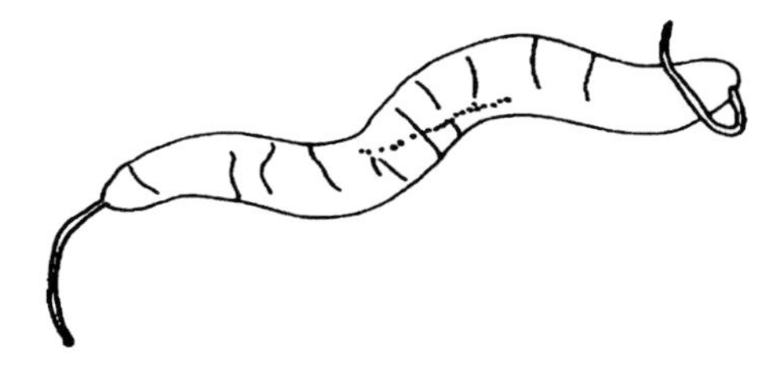

Figure 4.22 Schematic presentation of a magnetotactic bacterium (*Aquaspirillum magnetotacticum*) where the magnetosomes can be seen as black beads [119]

[22] Note that the earth's magnetic field has a strength of the order of 1 G, see [88].

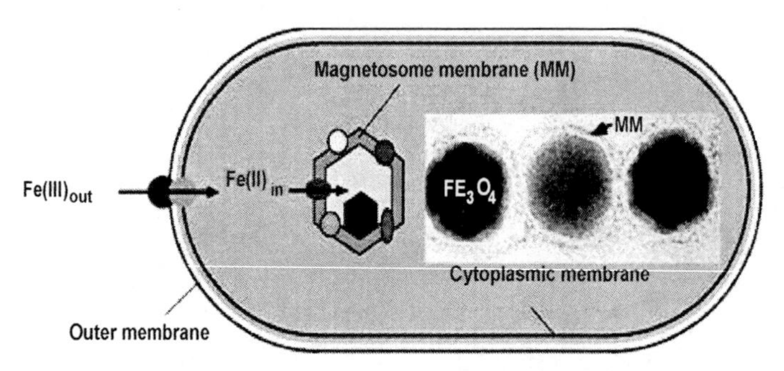

Figure 4.23 Possible mechanism of formation of magnetite within magnetosomes [122]

avoid the toxic effect of oxygen available in the surface water [88, 120]. These bacteria could be very important for the biogeochemical cycling of metals as when the bacteria die, sedimentation of fine magnetic particles will occur [118]; also, these bacteria have been reported to be useful for their potential ability to remove metals from contaminated soils [121].

But what does all this have to do with corrosion and MIC? There is some indirect and direct evidence here: magnetotactic cells can accumulate iron approximately 20,000- to 40,000-fold over its extracellular concentration [118]. Between 14% and 79% by weight of the magnetosome is magnetite (Fe_3O_4), where "the existence of … other oxides of iron or … iron sulphides in certain magnetotactic bacteria cannot be ruled out" [118]. If these bacteria need this much iron, from where can they get it?

One proposed model for magnetite biomineralisation in *Magnetospirillum* species is that Fe(III) is actively taken up by the cell, possibly via a reductive step, and then it is thought to be re-oxidised, resulting in magnetite production within the magnetosome, as seen in Figure 4.23 [122].

Could magnetosome formation mechanisms contribute to corrosion in the way that iron-reducing bacteria do by consuming ferric iron ions? While this is yet not known about magnetic bacteria, there is indirect evidence showing that the bacteria with magnetic properties could indeed be very important in MIC.

In their investigation, Bahaj *et al.* [123] used *Gallionella ferruginea* that are known to form tubercles and MIC [80] and accumulate iron hydroxide in their bodies. If these bacteria are present in an iron-rich medium, they pick up iron, and due to the increase of iron concentration in their bodies, their magnetic susceptibility and tendency for the attachment to magnetic surfaces such as iron also increase. This in turn will increase the likelihood of biofilm formation and hence further enhancement of corrosion. As those investigators put it, the "interaction" between the iron "in" the micro-organism and the iron "out" of the micro-organism, *i.e.*, the metallic substrate, could result from factors such as [123]:

- Existence of a magnetic substrate (steel surface, for instance)
- Magnetic features of corrosion products, including various iron oxides such as magnetite

- Formation of a wide range of (ferromagnetic) sulphides during MIC, induction of magnetic fields due to factors such as application of CP systems (especially impressed current), use of electric welding facilities and transportation means such as electric trains or trams

Bahaj *et al.* were able to establish a way to explain, at least theoretically, how magnetic fields may be effective in encouraging biofilm formation and MIC. Javaherdashti [119] proposed using magnetotactic bacteria to, literally, corral corrosion-enhancing bacteria at a suitable corner of a system and then expose them to MIC chemical (biocide application) or physical (filtration) mitigation methods.

Certainly, there are still many puzzles in dealing with magnetic bacteria. However, using these bacteria in mitigation programs may prove to be more efficient than other MIC control methods, if research in this very new and exotic area of MIC is supported in the manner it deserves.

4.14 Summary and Conclusions

Microbiologically influenced corrosion (MIC) is a subdivision of biocorrosion that deals with the role of micro-organisms such as bacteria in the initiation of and increasing both the intensity and extent of corrosion.

MIC is so important that its industrial, economic and even public health-related impact cannot be overlooked. MIC-related expenses account for a certain fraction of the GNP (about 0.8% GNP calculated), and the domain of its effects can be as far reaching as agriculture and even some diseases.

MIC is electrochemical in essence. However, it does not have a straightforward electrochemistry. For more than seven decades, researchers have been trying to explain MIC with electrochemistry, but it seems that the bacteria have more surprises in store for us: while the classic theory proposed in the mid-1930s put all the blame on SRB, the alternative, new theories tried to sequester the bacteria as much as possible. It was recently suggested that perhaps the bacteria themselves are engaged in picking up the required electrons directly from the metallic surface. However, these new findings need to be further examined to explore the complexities encountered in practice more efficiently.

SRB are not the only bacteria, or even the most important bacteria, involved in MIC. There are many bacteria that could be much more interesting than SRB. Although SRB and their corrosive effects and (for the first time) their impact on stress corrosion cracking were discussed here, another example of the bacteria involved in corrosion was also presented. This was a group of bacteria collectively named the iron-reducing bacteria (IRB).

IRB are interesting not only because of their possible corrosivity and their impact on accelerating of stress corrosion cracking processes (also discussed for the first time here), but also because of their possible protective and inhibitive features on corrosion.

Despite what we know about micro-organisms and their role in corrosion, we must be humble and honest to say that these tiny little living things do have the power of puzzling us. Comparing what we know about them with what we do not know is like comparing a single grain of sand with the beach.

It is very crucial to learn more about MIC and how it affects our industrial systems, because of the risks involved, both economic and environmental. Logically, in order to know more, much better conditions of research and development are required, and to achieve this, more funding are essential. To attract more funding, apart from considering economic and environmental risks, industry needs to know how systems can be become vulnerable to MIC, as prevention is much better than mitigation.

The next chapter deals with general guidelines for finding out how industrial systems – whether a heat exchanger, a gas pipeline or a ballast tank – can be in danger of being attacked by MIC.

References

1. Little BJ, Lee J, Ray R (2007) How marine conditions affect the severity of MIC of steels. MIC – An International Perspective Symposium; Extrin Corrosion Consultants, Curtin University, Perth, Australia, February 14–15, 2007
2. Franklin MJ, White DC, Isaacs H (1991) Pitting Corrosion by Bacteria on Carbon Steel, Determined by the Scanning Vibrating Electrode Technique, Corrosion Science, Vol. 32, No. 9, pp. 945–952
3. Sandoval-Jabalera R, Nevarez-Moorillon GV, Chacon-Nava JG, Malo-Tamayo JM, Martinez-Villafane A (2006) Electrochemical Behaviour of 1018, 304 and 800 Alloys in Synthetic Wasterwater, Journal of Mexican Chemical Society, vol. 50, No. 1, pp. 14–18
4. Sand W (1997) Microbial mechanisms of deterioration of inorganic substrates: A general mechanistic overview. International Biodeterioration & Biodegradation 40(2–4):183–190
5. de Romero MF, Urdaneta S, Barrientos M, Romero G (2004) Correlation Between Desulfovibrio Sessile Growth and OCP, Hydrogen Permeation, Corrosion Products and Morphological Attack on Iron, Paper No. 04576, CORROSION 2004, NCAE International
6. Beech I, Bergel A, Mollica A, Flemming H-C (Task Leader), Scotto V, Sand W (2000) Simple Methods for The Investigation of the Role of Biofilms in Corrosion, Brite Euram Thematic Network on MIC of Industrial Materials, Task Group 1, Biofilm Fundamentals, Brite Euram Thematic Network No. ERB BRRT-CT98-5084, September 2000
7. Little BJ, Wagner P (1997) Myths related to microbiologically influenced corrosion. Mater Perform 36(6):40–44
8. Li SY, Kim YG, Jeon KS, Kho YT, Kang T (2001) Microbiologically Influenced Corrosion of Carbon Steel Exposed to Anaerobic Soil, *CORROSION*, Vol. 57, No. 9, pp. 815–828, September 2001
9. Powell C (2006) Review of Splash Zone Corrosion and Biofouling of C70600 Sheathed Steel during 20 Years Exposure, Proceedings of EuroCorr 2006, 24–28 September 2006, Maastricht, the Netherlands
10. Palraj S, Venkatacahri G (2006) Corrosion and Biofouling Characteristics of Mild Steel in Mandapam Waters, *Materials Performance (MP)* vol. 45, No. 6, pp. 46–50
11. Javaherdashti R (1999) A review of some characteristics of MIC caused by sulphate-reducing bacteria: Past, present and future. Anti-Corrosion Methods & Materials Vol. 46, No. 3, pp. 173–180

12. Flemming HC (1996) Economical and technical overview. In: Microbially influenced corrosion of materials, Heitz E, Flemming HC, Sand W (eds), Springer-Verlag Berlin, Heidelberg
13. Javaherdashti R, Singh Raman RK (2001) Microbiologically influenced corrosion of stainless steels in marine environments: A materials engineering approach. In: Proceedings of engineering materials. The Institute of Materials Engineering, Australia, 23–26 September 2001
14 Singleton, R (1993) The sulphate-reducing bacteria: An overview. In: The sulfate-reducing bacteria: Contemporary perspectives, Odom JM, Singleton J-R (eds). Springer-Verlag, New York
15. Maxwell S, Devine C, Rooney F, Spark I (2004) Monitoring and Control of Bacterial Biofilms in Oilfield Water Handling Systems, Paper No. 04752, CORROSION 2004, NCAE International
16 Tributsch H, Rojas-Chapana JA, Bartels CC, Ennaoui A, Hofmann W (1998) Role of Transient Iron Sulfide Films in Microbial Corrosion of Steels, *CORROSION,* Vol. 54, No. 3, pp. 216–227, March 1998.
17. de Romero M, Duque Z, de Rincon O, Perez O, Auranjo J, Martinez A (2000) Online monitoring systems of microbiologically influenced corrosion on Cu–10% Ni alloy in chlorinated, brackish water. CORROSION 55(8):867–876
18 Cord-Ruwisch R (1996) MIC in hydrocarbon transportation systems. Corrosion Australasia 21(1):8–12
19. Li SY, Kim YG, Kho YT (2003) Corrosion behavior of carbon steel influenced by sulfate-reducing bacteria in soil environments, Paper No 03549, CORROSION 2003, NACE International
20. King RA (2007) Microbiologically induced corrosion and biofilm interactions. MIC – An International Perspective Symposium, Extrin Corrosion Consultants, Curtin University, Perth, Australia 14–15 February 2007
21. Javaherdashti R, Sarioglu F, Aksoz N (1997) Corrosion of drilling pipe steel in an environment containing sulphate-reducing bacteria. Intl J Pres Ves Piping (73):127–131
22. Angell P, Urbanic K (2000) Sulphate-reducing bacterial activity as a parameter to predict localized corrosion of stainless alloys. Corrosion Science (42):897–912
23. Javaherdashti R (2004) On the role of MIC in non-lethal biological war techniques. Proceedings of Weapons, Webs and Warfighters, Land Warfare Conference 2004, 27–30 September 2004, Melbourne, Australia
24. Walsh D, Pope D, Danford M, Huff T (1993) The Effect of Microstructure on Microbiologically Influenced Corrosion, *Journal of Materials (JOM),* Vol. 45, No. 9, pp. 22–30, September 1993
25. Stott JFD (1993) What progress in the understanding of microbially induced corrosion has been made in the last 25 years? (A Personal Viewpoint) Corrosion Science 35(s1–4):667–673
26. Fitzgerald III JH (1993) Evaluating Soil Corrosivity – Then and Now, Materials Performance (MP). Vol. 32, No. 10, pp. 17–19, October 1993
27. McKubre MCH, Syrett BC (1986) Harmonic Impedance Spectroscopy for the Determination of Corrosion rates in Cathodically Protected Systems, Corrosion Monitoring in Industrial Plants Using Nondestructive Testing and Electrochemical Methods, ASTM STP 908, G.C. Moran and P. Labine (Eds.), American Society for Testing and Materials, Philadelphia
28. Stott JFD, Skerry BS, King RA (1988) Laboratory evaluation of materials for resistance to anaerobic corrosion caused by sulphate reducing bacteria: Philosophy and practical design, the use of synthetic environments for corrosion testing. ASTM STP 970, Francis PE and Lee TS (eds), ASTM
29. Videla HA (2007) Mechanisms of MIC: Yesterday, today and tomorrow. MIC – An International Perspective Symposium, Extrin Corrosion Consultants, Curtin University, Perth, Australia, 14–15 February 2007
30. Geesey GG (1993) Biofilm Formation, In: A Practical Manual on Microbiologically-influenced CorrosionKobrin G (ed), NACE, Houston, Texas, USA

31. Obuekwe CO, Westlake DW, Plambeck JA, Cook FD (1981) Corrosion of mild steelin cultures of ferric iron reducing bacterium isolated from crude oil, polarisation characteristics. CORROSION 37(8):461–467
32. Little BJ, Wagner P, Hart K, Ray R, Lavoi D, Nealson K, Aguilar C (1997) The role of metal reducing bacteria in microbiologically influenced corrosion, Paper No. 215. CORROSION/97, NACE, Houston, Texas USA
33. Dexter SC, LaFontain JP (1998) Effect of natural marine biofilms on galvanic corrosion. CORROSION 54(11):851–861
34. Guiamet PS, Gomez de Saravia SG, Videla HA (1999) An innovative method for preventing biocorrosion through microbial adhesion inhibition. International Biodeterioration & Biodegradation (43):31–35
35. Al-Hashem A, Carew J, Al-Borno A (2004) Screening test for six dual biocide regimes against planktonic and sessile populations of bacteria. Paper No. 04748, CORROSION 2004, NACE International
36. Xu K, Dexter SC, Luther GW (1998) Voltametric microelectrodes for biocorrosion studies. CORROSION 54(10):814–823
37. Liu H, Xu L, Zeng J (2000) Role of corrosion products in biofilms in microbiologically induced corrosion of carbon steel. Brit Corrosion J 35(2):131–135
38. Taheri R, Nouhi A, Hamedi J, Javaherdashti R (2005) Comparison of corrosion rates of some steels in batch and semi-continuous cultures of sulfate-reducing bacteria. Asian J Microbiol Biotech Env Sc 7(1):5–8
39. Dickinson WH, Lewandowski Z, Geer RD (1996) Evidence for surface changes during ennoblement of type 316 stainless steel: dissolved oxidant and capacitance measurements. CORROSION 52(12):910–920
40. Videla HA (1996) Manual of Biocorrosion. Ch. 4, CRC Press
41. Dexter SC, Chandrasekaran P (2000) Direct Measurement of pH within Marine Biofilms on Passive Metals, *Biofouling*, Vol. 15, No. 4, pp. 313–325
42. Landoulsi J, Pulvin S, Richard C, Sabot K (2006) Biocorrosion of Stainless Steel in Artificial Fresh Water: Role of Enzymatic Reactions, Proceedings of EuroCorr 2006, 24–28 September 2006, Maastricht, the Netherlands.
43. Scotto V, Mollica A (2000) A Guide to Laboratory Techniques for the Assessment of MIC Risk due to the Presence of Biofilms. www.corr-institute.se/english/Web_DT/files/-MICbook.pdf, September 2000
44. Kovach CW, Redmond JD (1997) High performance stainless steels and microbiologically influenced corrosion. www.avestasheffield.com, acom 1–1997
45. Salvarezza RC, Videla HA (1980) *CORROSION*, Vol. 36, pp. 550–554
46. Pope DH, Morris III EA (1995) Some experiences with microbiologically-influenced corrosion, Mater Perform, 34(5):23–28
47. Borenstein SW, Lindsay PB (1987) MIC failure analyses. Paper No. 381, Corrosion/87, NACE, Houston, Texas, USA
48. Metals Handbook, Vol. 13, Corrosion, 9th edition, ASM, Metals Park, USA, p. 122
49. Wilderer PA, Characklis WG (1989) Structure and function of biofilms. In: Characklis WG and Wilderer PA (eds), Structure and Function of Biofilms, pp. 5–17, John Wiley and Sons, New York
50. A Working Party Report on Microbiological Degradation of Materials – And Methods of Pretection (1992) Section 433, European Federation of Corrosion Publications, No. 9, The Institute of Materials, London
51. Roe FL, Lewandowski Z, Funk T (1996) Simulating microbiologically influenced corrosion by depositing extracellular biopolymers on mild steel. CORROSION, 52(10):744–752
52. Lewandowski Z, Funk T, Roe FL, Little BJ (1994) Spatial Distribution of pH at Mild Steel Surfaces using an Iridium Oxide Microelectrode, in: Microbiologically Influenced Corrosion Testing, Kearns, JR, Little BJ (eds), STP 1232, ASTM, 1994, USA
53. Kearns JR, Little BJ(Eds.) (1994) STP 1232, ASTM, USA.

54. Chan G, Kagwade SV, French GE, Ford TE, Mitchell R, Clayton CR (1996) Metal Ion and Exopolymer Interaction: A Surface Analytical Study, *CORROSION* Vol. 42, No. 12, pp. 891–899
55. Lewandowski Z, Stoodley P, Altobelli S (1995) Experimental and conceptual studies on mass transport in biofilms. Water Sci Technol (31):153–162
56. Hernandez G, Kucera V, Thierry D, Pedersen A, Hermansson M (1994) Corrosion Inhibition of Steel by Bacteria, CORROSION, 50(8):603–608
57. Jack RF, Ringelberg DB, White DC (1992) Differential corrosion rates of carbon steel by combinations of *Bacillus* sp, *Hania alvei* and *Desulfovibrio gigas* established by phospholipid analysis of electrode biofilms. Corrosion Science 33(12):1843–1853
58. Graff WJ (1981) Introduction to offshore structures. Ch. 12, Gulf Pub Co., Houston, Texas, USA
59. Obuekwe CO, Westlake DWS, Cook FD, Costerton JW (1981) Surface changes in mild steel coupons from the action of corrosion-causing bacteria. Appl Envir Microbiol 41(3):766–774
60. Borenstein SW (1988) Microbiologically influenced corrosion failures of austenitic stainless steel welds. Mater Perform 27(8):62–66
61. Stoecker JG (1993) Penetration of stainless steel following hydrostatic test. In: A practical manual on microbiologically influenced corrosion. Kobrin G (ed), NACE, Houston, Texas, USA
62. Ornek D, Wood TK, Hsu CH, Sun Z, Mansfeld F (2002) Pitting corrosion control of aluminum 2024 using protective biofilms that secrete corrosion inhibitors. CORROSION 58(9):761–767
63. Nagiub A, Mansfeld F (2002) Microbiologically influenced corrosion inhibition observed in the presence of *Shewanella* micro-organisms. Proceedings of 15th International Corrosion Council, Spain
64. Dubiel M, Hsu CH, Chien CC et al (2002) Microbial iron respiration can protect steel from corrosion. Appl Environ Microbiol 68(3):1440–1445
65. Obuekwe CO, Westlake DWS, Cook FD (1981) Effect of nitrate on reduction of ferric iron by a bacterium isolated from crude oil. CANJMICROBIOL (27):692–697
66. Lee AK, Newman DK (2003) Microbial iron respiration: impacts on corrosion processes, online, Appl Environmental Microbiol, No. 7
67. Videla HA (1996) Manual of Biocorrosion, pp. 74–120 and 193–196, CRC press, Inc.
68. Newman RC, Rumash K, Webster BJ (1992) The effect of pre-corrosion on the corrosion rate of steel in natural solutions containing sulphide: Relevance to microbially influenced corrosion. Corrosion Science 33(12):1877–1884
69. Byars HG, (1999) Corrosion Control in Petroleum Production, Chapter 2, TPC Publications 5, second edition, NACE international
70. Archer ED, Brook R, Edyvean RGJ, Videla HA (2001) Selection of Steels for use in SRB Environments, Paper No. 01261, Corrosion 2001, NACE International
71. Sanchez del Junco A, Moreno DA, Ranninger C, Ortega-Calvo JJ, Saiz-Jimenez C (1992) Microbial induced corrosion of metallic antiquities and works of art: A critical review. International Biodeterioration & Biodegradation (29):367–375
72. Hamilton WA (1985) Sulphate-reducing bacteria and anaerobic corrosion. Ann Rev Microbiol 39):195–217
73. Chamritski IG, Burns GR, Webster BJ, Laycock NJ (2004) Effect of iron-oxidizing bacteria on pitting od stainless steels. CORROSION 60(7)
74. Critchley M, Javaherdashti R (2005) Materials, micro-organisms and microbial corrosion – A review. Corrosion and Materials 30(3):8–11
75. Jones DA, Amy PS (2002) A thermodynamic Interpretation of Microbiologically Influenced Corrosion, CORROSION, Vol. 58, No. 8, pp. 638–645, August 2002
76. Jack TR (2002) Biological Corrosion Failures, ASM International, March 2002
77. Blackburn FE (2004) Non-bioassy Techniques for Monitoring MIC, Corrosion 2004, paper 04580, NACE International

78. Marconnet C, Dagbert C, Roy M, Feron D (2006) Microbially Influenced Corrosion of Stainless Steels in the Seine River, Proceedings of EuroCorr 2006, 24–28 September 2006, Maastricht, the Netherlands
79. Barton LL, Tomei FA (1995) Characteristics and activities of sulfate-reducing bacteria in sulfate-reducing bacteria, Barton LL (ed), Biotechnology Handbooks, Vol. 8, Plenum Press, New York
80. Stott JFD (1988) Assessment and control of microbially induced corrosion. Metals and Materials 224–229
81. King RA (2007) Trends and developments in microbiologically induced corrosion in the oil and gas industry. MIC – An International Perspective Symposium, Extrin Corrosion Consultants, Curtin University, Perth, Australia, 14–15 February 2007
82. Miller JDA, Tiller AK (1970) Microbial aspects of metallurgy, Miller JDA (ed), American Elsevier Publishing, New York
83. The Role of Bacteria in the Corrosion of Oilfield Equipment (1982) TPC3, NACE International
84. Willis CL, Gibson GR, Holt J, Allison C (1999) Negative correlation between oral malodour and numbers and activities of sulphate-reducing bacteria in the human mouth. Arch Oral Biol (44):665–670
85. Langendijk PS, Hagemann J, Van der Hoeven JS (1999) Sulfate-reducing Bacteria in Periodontal Pockets and in Healthy Oral Sites, J. Clin. Periodonotl., Vol. 26, pp. 596–599
86. McDougall R, Robson J, Paterson D, Tee W (1997) Bacteremia Caused by a Recently Described Novel Desulfovibrio Species, *Journal of Clinical Microbiology*, pp. 1805–1808, July 1997
87. Lfill, C (1999) The isolation and Purification of Sulphate-reducing Bacteria from the Colon of Patients Suffering from Ulcerative Colitis, B.Sc. (Hons) School of Pharmacy and Biomedical Sciences, University of Portsmouth, UK, June 1999.
88. Blakemore RP, Frankel RB (1981) Magnetic navigation in bacteria. Sci Amer (245):42–49
89. Javaherdashti R (2005) Microbiologically influenced corrosion and cracking of mild and stainless steels. PhD Thesis, Monash University, Australia
90. Tiller AK (1983) Electrochemical aspects of microbial corrosion: An overview proceedings of microbial corrosion, 8–10 March 1983, The Metals Society, London
91. Rainha VL, Fonseca ITE (1997) Kinetics studies on the SRB influenced corrosion of steel: A first approach. Corrosion Science 39(4):807–813
92. Iverson WP (1998) Possible source of a phosphorus compound produced by sulfate-reducing bacteria that cause anaerobic corrosion of iron. Mater Perform 37(5):46–49
93. Videla HA, Herrera LK, Edyvean RG (2005) An updated overview of SRB induced corrosion and protection of carbon steel. Paper No. 05488, Corrosion 2005, NACE International
94. Smith JS and Miller JDA, (1975) Nature of Sulphides and Their Corrosive Effect on Ferrous Metals: A Review, British Corrosion Journal, Vol. 10, No. 3, pp. 136–143
95. Hang DT (2003) Microbiological study of the anaerobic corrosion of iron. PhD Dissertation, University of Bremen, Bremen, Germany
96. Scott PJB, Goldie J (1991) Ranking alloys for susceptibility to MIC – A preliminary report on high-Mo alloys, Mater Perform 30(1):55–57
97. Schutz RW (1991) A case for titanium's resistance to microbiologically influenced corrosion. Mater Perform 30(1):58–61
98. Wagner P, Little BJ (1993) impact of alloying on microbiologically influenced corrosion – A review. Mater Perform 32(9):65–68
99. Hardy JA, Brown JL (1984) The corrosion of mild steel by biogenic sulfide films exposed to air. CORROSION 40(12):650–654
100. Lee W, Characklis WG (1993) Corrosion of mild steel under anaerobic biofilm. CORROSION 49(3):186–198
101. Tiller AK (1983) is stainless steel susceptible to microbial corrosion? Proceedings of Microbial Corrosion, 8–10 March 1983, The Metals Society, London

102. Neville, A, Hodgkiess T (1998) Comparative study of stainless steel and related alloy corrosion in natural sea water. Brit Corrosion J 33(2):111–119
103. Johnsen R, Bardal E (1985) Cathodic properties of different stainless steels in natural seawater. CORROSION 41(5):296–302
104. Linhardt P (1996) Failure of chromium-nickel steel in a hydroelectric power plant by manganese-oxidising bacteria. In: Microbially influenced corrosion of materials. Heitz E, Flemming H-C, Sand W (eds), Springer-Verlag Berlin, Heidelberg
105. Stainless Steel Selection Guide (2002) Central States Industrial Equipment & Service, Inc., http://www.al6xn.com/litreq.htm, USA
106. Javaherdashti R, Raman Singh RK, Panter C, Pereloma EV (2006) Microbiologically assisted stress corrosion cracking of carbon steel in mixed and pure cultures of sulfate reducing bacteria. Intl Biodeterioration & Biodegradation 58(1):27–35
107. Javaherdashti R, Raman Singh RK, Panter C, Pereloma EV (2005) Role of microbiological environment in chloride stress corrosion cracking of steels. Materi Sci Technol 21(9):1094–1098
108. Javaherdashti R, Raman Singh RK, Panter C, Pereloma EV (2004) Stress corrosion cracking of duplex stainless steel in mixed marine cultures containing sulphate reducing bacteria. Proceedings of Corrosion and Prevention 2004 (CAP04), 21–24 November 2004, Perth, Australia
109. Singh Raman RK, Javaherdashti R, Panter C, Cherry BW, Pereloma EV (2003) Microbiological environment assisted stress corrosion cracking of mild steel Proceedings of Corrosion Control and NDT, 23–26 November 2003, Melbourne, Australia
110. Standard Test Method for Iron Bacteria in Water and Water-formed Deposits (1997) ASTM D932–85 (Reapproved 1997), ASTM annual book, ASTM, USA
111. Simpson WJ (1999) Isolation and characterisation of thermophilic anaerobies from bass strait oil production waters. MA:Sci Thesis, School of Applied Sciences, Monash University
112. Panter C (2007) Ecology and characteristics of iron reducing bacteria-suspected agents in corrosion of steels. MIC – An International Perspective Symposium, Extrin Corrosion Consultants, Curtin University, Perth, Australia, 14–15 February 2007
113. Panter C (1968) Iron reducing bacteria of soil. MSc Thesis, Dept. of Soil Science, University of Alberta, Canada
114. Kajiyama F, Okamura K (1999) Evaluating cathodic protection reliability on steel pipes in microbially active soils. CORROSION 55(1):74–80
115. Pope DH, Zintel TP, Aldrich H, Duquette D (1990) Efficacy of biocides and corrosion inhibition in the control of microbiologically influenced corrosion. Mater Perform 29(12):49–55
116. Lee AK, Buehler MG, Newman DK (2006) Influence of a dual-species biofilm on the corrosion of mild steel. Corrosion Science 48(1):165–178
117. Sakaguchi T, Tsujimura N, Matsunaga T (1996) A novel method for isolation of magnetic bacteria without magnetic collection using magnetotaxis. J Microbiol Meth (26):139–145
118. Hughes MN, Poole PK (1989) Metals and micro-organisms. Section 59, Chapman and Hall, New York
119. Javaherdashti R (1997) Magnetic bacteria against MIC. Paper No. 419, CORROSION 97, NACE International
120. Bean CP (1974) Magnetism and life in fundamentals of physics, Section E 14–1, by Halliday D, Resnick R, 3rd edn., c1990
121. Magnetic Bacteria May Remove Metals from Contaminated Soils (1997) Chemical News, Mater Perform 36(1):47
122. The Magneto-Lab, Dr. Dirk Schüler, Junior Group at the MPI for Marine Microbiology, Bremen, http://magnum.mpi-bremen.de/magneto/research/index.html
123. Bahaj AS, Campbell SA, Walsh FC, Stott JFD (1992) The importance of environmental factors in microbially influenced corrosion Part 2, Magnetic field effects in microbial corrosion. Proceedings of the Second EFC Workshop, Portugal 1991 (Sequeira CAC, Tillere AK, eds), European Federation of Corrosion Publications, No. 8, The Institute of Materials

Chapter 5
How Does a System Become Vulnerable to MIC?

5.1 Introduction

The late David White has been quoted as saying "Microbiologically influenced corrosion is industrial venereal disease: it's expensive, everybody has it and nobody wants to talk about it". If you ask any one who has been involved in MIC assessment for an industry, he can tell you stories for hours about how he has tried first to convince the industry about (1) the importance of corrosion treatment, and (2) the involvement of some "bugs" in corrosion. I personally believe that what happened in Alaska's Prudhoe Bay is an alert for all people who are involved in design, operation and maintenance: *do not underestimate bugs!*

This chapter will deal mainly with the problem of MIC recognition. In other words, we want to know what factors can be taken as indicators of MIC, principally independent of the system itself. So, what we are presenting in this chapter can be applied to systems such as pipelines, cooling systems, ballast tanks, hydrants and the like.

5.2 General Points Regarding the Vulnerability of Industrial Systems

What do we mean by an industrial system? An industrial system is a part of an industry that does a definite job within that industry and due to its working or service conditions, could be vulnerable to MIC. For instance, a pipeline that carries oil or gas is an example of industrial system as much as a ballast tank in a ship or a condenser in a power plant. It is very interesting to see that although industrial systems may differ from each other in shape, design and function, they will be affected by MIC under almost the same operating conditions. This is not only true for different systems, but also for diverse industries ranging from the oil industry to the power, mining, and ship industries and even the agricultural industry.

What will be addressed in this chapter regarding the risk of MIC and the vulnerability of a system are just "necessary" and "not enough" conditions. In other words, establishment of the following conditions can flag the danger of MIC but does not guarantee its occurrence.

An example of "necessary" but "not enough" factors could be what is known as a ranking table for the estimation and assessment of steel corrosion in marine moods: in the late 1970s, King proposed a "ranking table" with which, using the assessment of factors such as flow rate, oxygen, heavy metals and nitrogen and phosphorous contents, an index for the corrosivity of the seabed sediments could be produced [1]. In the early 1980s, Farinha found this ranking table inadequate to explain the marine muds in the UK [2]. One of the main drawbacks of King's ranking table is that it is designed for marine sediments, particularly open sea sediments, which are sufficiently different from estuarial or near shore harbour sediments [3]. While Farinha's index can be taken as a modification to King's method [4], among the factors that was added to his ranking table was the very important factor of sulphate concentration. Farinha's index has been used successfully in other studies [2]. However, from an MIC point of view, Farinha's model just considers the impact of SRB on corrosion and not other corrosion-related microbial species. Perhaps another ranking method that considers other types of bacteria can also be applied [5].

Another example is the very useful flow diagram proposed by Krooneman *et al.* for the assessment and reducing the risk of MIC in pipelines [6]. In this simple yet very clever flow diagram, the possibility of different factors leading to the risk of MIC is addressed. Some of these factors are oxygen content, pH, sulphate content, total organic carbon content, salt concentration and temperature. While the model is certainly a useful tool, it does not consider, for example, the possible effect of thermophilic corrosive bacteria, or the risk of corrosion when the temperature is above 40°C. This model may not be useful in assessing MIC in, say, geo-thermal power plants [7].

The above examples may serve to show that a better understanding of corrosion mechanisms and other involved factors can advance our understanding and power of corrosion prediction tremendously. It can also serve to emphasise that no matter how you do it, there are always factors in your list that may prevent it from being applicable everywhere and anytime. Therefore what will be discussed later in this chapter must be understood within the context of a "being-useful-so-far" basis.

As always reminded, it is of great importance to be able to distinguish MIC from other types of corrosion and to find the best remedy for it; otherwise, the problem will be increased.

5.3 Important Systems/Working Conditions Leading to MIC

By referring to the metallic substrate and the bulk water the "environment" (Figure 5.1), the following classification can be used [8].

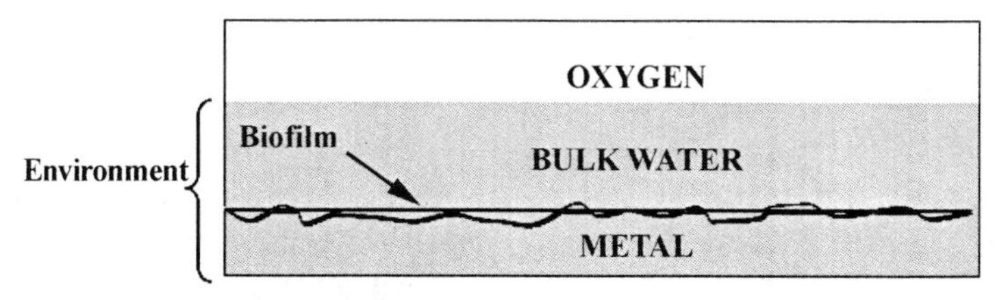

Figure 5.1 Components of the "environment"

The water's temperature, pH, and chemistry (chlorides, nutrients) are important factors to be reported, in addition to factors such as total dissolved solids (TDS) and suspended solids content [9]. To report such a diverse range of properties, a distinction can be made as follows: physico-mechanical effects and features of the environment including those of substrate (such as surface energy, roughness, surface temperature, residual stresses, *etc.*) and those of the bulk water (such as its mean linear velocity, nutrient concentration, temperature, and the like), and chemical effects and features of the environment including those of the substrate metal (such as the existence and/or absence of some alloying elements that can encourage the growth/attachment of the bacteria) and those of the bulk water (such as existing ions, TDS, *etc.*). These two sets of effects are both totally arbitrary and interrelated to each other.

5.4 Physico-mechanical Factors and Their Effects on MIC

Figure 5.2 is a schematic representation of the relationship among pH, flow velocity and temperature with regard to the observed corrosion rate (pH has been considered a physical rather than chemical factor with regard to the other two features that contribute to MIC).

What is emphasised in Figure 5.2 is that micro-organisms, like human beings, are capable of living only within a certain range of temperature and pH. If the measured corrosion rate is only high between two temperatures T_1 and T_2 and then it decreases at temperatures below T_1 and above T_2, then chances are that the type of corrosion could be MIC-related and not, for example, high-temperature corrosion, as illustrated in Figure 5.3.

The same is also true with pH. There is a range of pH that is tolerable for bacteria, and although there may be some exceptions to this rule, a certain range of pH can always be defined for a certain group of bacteria. For example, the pH range that is tolerable to acid-loving bacteria is not suitable for SRB. Flow velocity is also an important factor, though this may not be true in all cases. Kobrin [10] reported that in stainless steel tubes in which the water velocity was higher than

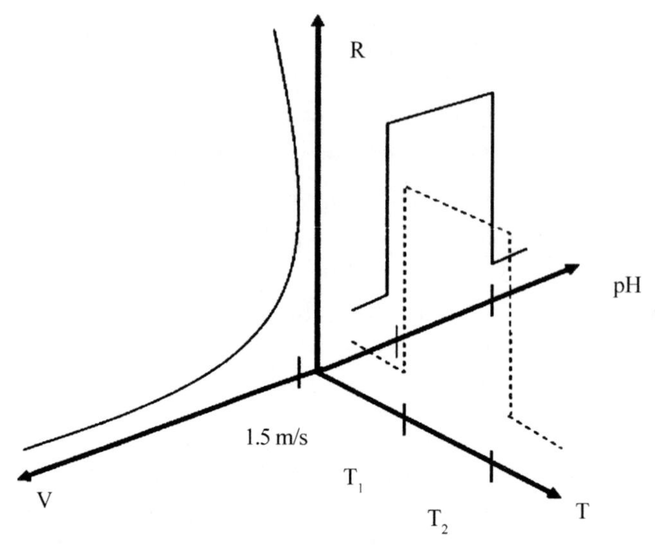

Figure 5.2 Schematic relationship between flow velocity (V), pH of the bulk solution, and temperature (T) with respect to the observed corrosion rate (R)

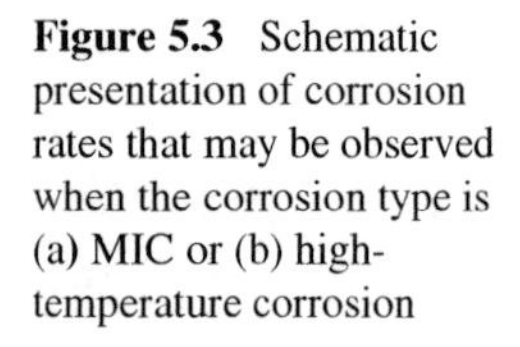
Figure 5.3 Schematic presentation of corrosion rates that may be observed when the corrosion type is (a) MIC or (b) high-temperature corrosion

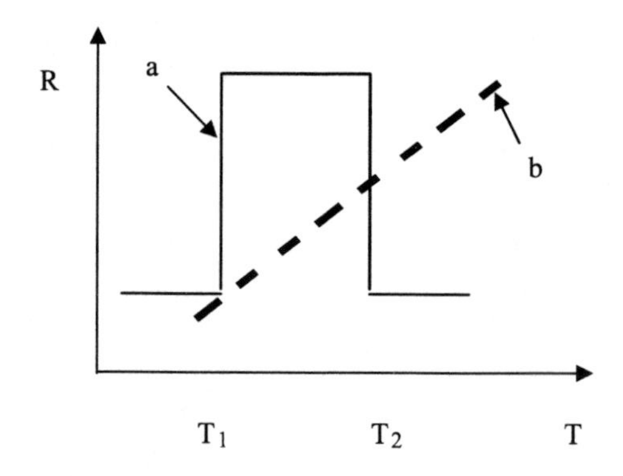

the normally accepted 1.5 m/s criterion,[1] MIC was still operative. However, it is a trend that by slowing down the fluid flow velocity, perhaps due to the absence of mechanical sheer forces that would otherwise interrupt biofilm formation, the likelihood of MIC also increases.

---

[1] It is generally recommended to keep water flow velocity more than 1.5 m/s, pH above 10–11 and temperature well above 90°C to lower the risk of MIC. Reader should understand that beyond these seemingly rigid rules and regulations, there are huge uncertainties, making them be understood as a whole not as isolated items. For example, you may try to keep water flowing and still the probability of getting no MIC may not be nil.

5.4.1 Chemical Factors

A very important factor in a system that can lead to MIC is the water quality: if raw, untreated or poorly treated water is being used for an industrial activity such as hydrotesting, one may expect that the risk of MIC will be very high.

What is meant by untreated water is water in which no specific physical/chemical treatment has been done to remove (mainly) corrosion-related bacteria. This water can be sea, river, or well water used for industrial activities.

5.4.2 Water Treatment

The water treatment method depends on many variables including, but not limited to, the availability of alternative methods, the economics of replacements, and practical limitations related to the implementation of a physical/chemical treatment.

A crucial aspect of water is its total dissolved solids (TDS). The existence of halophilic (salt-loving) SRB in waters with very high TDS (240,000 mg/L) has been reported [11]. Other aspects of water/biocide interaction (such as water activity and TDS content) will be discussed in more detail in Chapter 9.

5.4.3 Oxygen

The oxygen concentration of water, as bulk fluid, may not always be useful and, in fact, it can even be deceiving. As discussed in Chapter 4, biofilms are capable of forming anaerobic patches in otherwise aerobic bulk solutions. It has been reported that a biofilm with a thickness of only 12 μm may be sufficient to create totally anaerobic regions in an anaerobic system where at the base of the biofilm, SRB can be motile and active [11]. Having said that, some researchers believe that by knowing the chemical oxygen demand (COD), it may be possible to know the concentration of electron donors available for sulphate or metal reduction so that a low COD would mean a low risk of availability of SRB or other types of "reducers" such as IRB [12].

5.4.4 Nutrients and the Ease of Reaching Them

The availability of nutrients is also a crucial factor, as it may be the principle factor in determining whether the prevailing bacterial population will be planktonic or sessile. Enos and Taylor reported on how the nutrient level could affect the "mode" of bacterial spatial position: when the environment is poor in nutri-

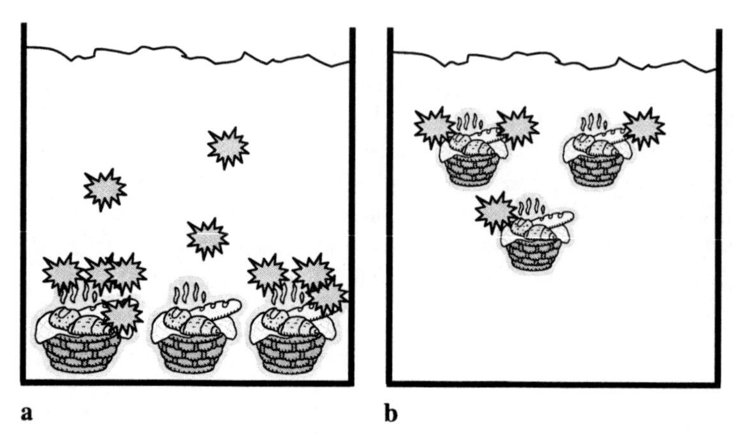

Figure 5.4 Relationship between nutrient level and bacterial attachment. **a** A nutrient-poor environment where the nutrients can be reached on the surface, thus promoting sessile growth. High corrosion rates can be expected. **b** A nutrient-rich environment where the nutrients can be reached within the bulk solution, thus promoting planktonic growth. Lower corrosion rates can be expected

ents, the required nutrients are precipitated onto the surface, and therefore the bacteria will settle down on the "food" that is available on the surface [13]. These "seated" bacteria are called the "sessile bacteria". In contrast, if the environment is rich in nutrients, the bacteria do not need to go to the surface, so planktonic (floating) growth is favoured. These situations are schematically presented in Figure 5.4.

Perhaps one immediate important outcome of such a scenario, as Enos and Taylor have also put it, is how reliable the microbial laboratory test results are (as they are basically run in rich-nutrient culture media) compared to real-life experiences (where the nutrient level may not always be that high). In other words, when under laboratory conditions the bacteria have all the required nutrients around, they may be inclined to prefer planktonic growth over sessile conditions, whereas a great majority of MIC problems come from sessile bacteria not planktonic ones (this issue will be discussed later in this chapter). While this concern is understandable and quite valid, one should not forget the intrinsic limitations of doing microbial corrosion experimentation, as follows.

Conducting a precisely controlled test that involves all the bacteria and determination of the share of each species in the evaluation and assessment of corrosion is impractical and almost impossible. Moreover, it is not practically possible to conduct a test where all the incorporating values and factors are exactly replicas of the natural conditions outside the walls of the laboratory. The following may illustrate how the real-life conditions and laboratory conditions may not be compatible.

Due to practical limits such as difficulty in establishing and running continuous and/or semi-continuous cultures, some MIC experimenters do the tests in batch-type cultures. In these cultures, a certain amount of food (nutrients) is provided to the bacteria, and no further change or displacement of the quantity of the nutrients is made. This, then, will be in contrast to continuous and/or semi-continuous cultures where the culture/nutrient quantities are frequently or continuously changed.

Although it is true that batch-type experiments may not be able to present the natural habitat of the bacteria as closely as continuous cultures (because in a semi-continuous or continuous test regime the supply and demand of the nutrients is more or less similar to those in nature [14, 15][2]), it can be argued for batch cultures that they can be the most similar and closest to mimicking what is happening in a stagnant water environment. When one considers a stagnant water environment, one can easily see that due to very slow movement (even almost no movement), no exchange of nutrients in that portion of the system will take place, so that the bacteria will have to feed on what is available in that particular environment. Therefore, it follows that the batch cultures may be better representatives of some natural conditions than are semi-continuous and continuous test regimes. Yet, the batch cultures are far from the ideal simulation of stagnant environments, obviously due to factors such as the kinetic discrepancies of the two environments.

It thus can be concluded that the reliability of laboratory test results is not 100%, and that such results must be used with care, keeping in mind all of their advantages and disadvantages.

One of the possible effects of sessile bacteria on planktonic bacteria is actually addressing the effect that biofilms can have on accelerating biocorrosion. This discussed was in Chapter 4 and will not be repeated here.

5.4.5 Alloying Elements and Their Impacts

Alloying elements are added to improve mechanical and electrochemical properties of the metal. For example, it is a well-known practice to add chromium to steel to increase its corrosion resistance. However, alloying elements can sometimes have other impacts as well; they may affect the way the metal responds to the environment from a microbial corrosion point of view.

For instance, it has been reported [16] that by increasing the sulphur content as an alloying element, the likelihood of tubercle formation also increases, and that molybdenum can reduce bacterial viability [17]. Lopes *et al.* investigated the factors that can help adhesion of *Desulfovibrion desulfuricans* on metallic and non-metallic surfaces [18]. They showed that adhesion of this group of SRB on nickel surfaces is relatively more significant compared to stainless steel 304 or polymethylmethacrylate (PMMA) surfaces, implying that the bacteria did show a powerful tendency to colonise on nickel surfaces.

---

[2] See also [15] Sections 2.5.10 and 2.5.11 discuss about advantages and disadvantages of continuos cultures over culture methods that could be very instructive.

The above shows just a few examples of possible enhanced interactions between the bacteria and some alloying elements. It is still not well known what the real mechanism(s) behind such behaviour could be. Whether such behaviour is the result of some sort of chemical response, the production of "adhesion proteins" [18], or any other mechanism(s), the end result is that some alloying elements do have some impact on MIC that will make the use of the material containing those alloying elements a matter of caution where the risk of MIC is involved.

5.4.6 Welding

Another important factor that has a very crucial impact on rendering a system vulnerable to MIC is welding. For engineering applications, welding is one of the most frequently applied methods used for "adhering" metallic parts to each other. However, no matter how useful it is, welding must be regarded as equivalent to a wound in a body: it always requires the highest attention, and it can be the most likely spot for the initiation of problems (compare infections in the human body and weld decay in welded structures).

According to Kurissery *et al.* [19], the first study reporting weldments as preferred spots for microbial colonisation dates back to 1950. Those researchers also quote some references where most of the corrosion failures in cooling water systems made up of "corrosion-resistant alloys" are around or within weldments.

When a piece of steel is welded to another, both the temperature and grain boundary energy distribution change along side the welding area (Figure 5.5).

As it is seen from Figure 5.5, at and around the weld zone, both the temperature and energy distribution curves show peaks. This can be interpreted as one of the

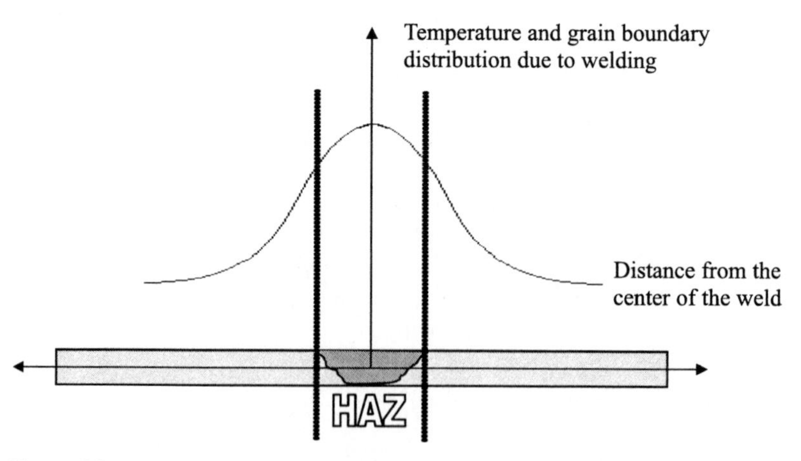

Figure 5.5 Schematic distribution of temperature due to welding and grain boundary energy in a metallic bar. The dark area between the dashed lines presents the heat-affected zone (HAZ)

main features of welding,[3] introducing zones where thermodynamically the energy is high and the structure has lost its uniform texture. The change in texture and thus surface roughness has been reported as a very important factor in the initiation of bacterial attachment [19, 20]. The possible effects of welding on accelerating MIC on metallic surfaces can be short-listed as follows [13, 16, 19–23]:

- Change of the surface roughness so that bacterial colonisation can be facilitated by "hooking" onto the rough surfaces.
- Change of the surface chemistry and microstructure of both the fusion and the HAZ and facilitation of the segregation of alloying elements, therefore making the surface more receptive in terms of bringing the alloying elements from within the bulk materials onto the surface, thus letting the micro-organisms have a better chance of finding the required nutrients.
- Welding can result in generating a heterogeneous surface so that formation of electrochemical cells on the surface may become much easier.
- Introducing/highlighting the impact of factors such as the existence of inclusions, and secondary phases.

As it appears, if post-welding treatments (such as stress relief, trimming, and finishing the welded area) are ignored, chances are that welding through a series of changes that are introduced into the parent material will promote the possibility of colonisation by sessile bacteria in terms of biofilm formation, and microbial corrosion gets started. *So, treat weldings as you would treat wounds to your own body.*

5.4.7 Impact of Hydrotesting

Another very important factor in the initiation of MIC is wrong or incomplete hydrotesting. While pneumatic testing is just leakage testing, hydrotesting is both leak and strength testing [24].

Hydrotesting is a routine test in industry to assess mainly the strength of weldments in systems that will operate under pressure. The test is done by introducing water into the system and applying internal pressures about 1.1 times the pressure that the system will undergo in real practice [25]. To carry out the hydrotest, most of the time untreated water (well water, river water or seawater) is used that may carry corrosion-related bacteria such as sulphate-reducing bacteria (SRB) or iron bacteria (IB). Figure 5.6 shows the pitting in the interior side of a tank after the tank was hydrotested with untreated water and the water was leftt in the tank for weeks. As can be seen from Figure 5.6, very severe pitting has occurred all over the surface.

[3] Kurissery *et al.* [19] quote from two references (references 22 and 23 in their papers) to explain how grain boundary energy content can affect bacterial attachment. In their reasoning, as bacteria are themselves negatively-charged, "chances are more for the cells to be attracted towards the grain boundaries with a high energy level and elemental segregation".

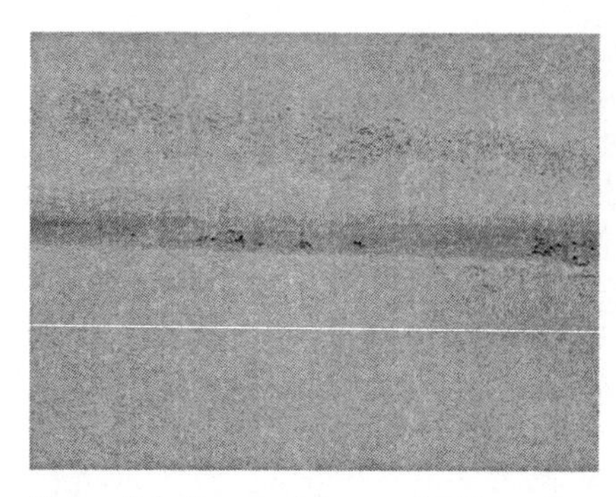
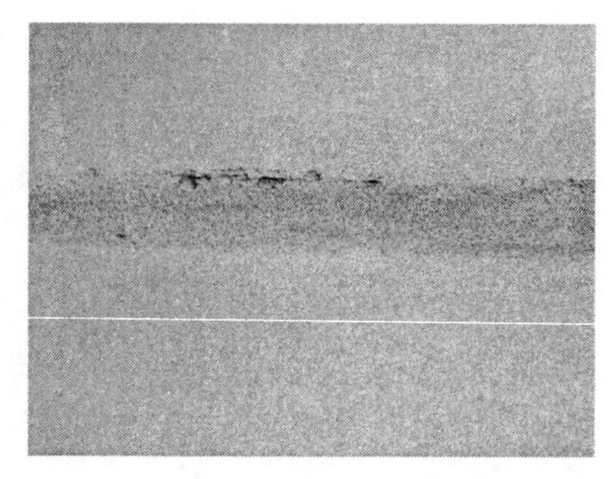

Figure 5.6 Severe pitting resulted from the water left in a vessel after hydrotesting, which is an example of wrong/inadequate hydrotesting. The pitting varied from 1.5 to 2.0 mm in depth and was about 2–5 mm in diameter. (courtesy of Extrin Consultants)

To gain a better understanding of the possible relationship between hydrotest and MIC, it is worth distinguishing between two types of hydrotest implementation methods, defined as follows [25]:

1. Wrong hydrotest: Untreated water has been used for the test. The untreated water is water for which no chemical treatment (mainly by biocides) to remove corrosion-enhancing bacteria has been done.
2. Incomplete hydrotest: Operations such as draining and drying immediately to be done after the test either have not been carried out or were poorly done so that, as evidence, one can still see water in the system.

As said above, normally untreated water is used for hydrotesting. This is perhaps due to two contradicting ideas about hydrotesting and its importance:

1. *Hydrotesting is important* because it measures how strong the component is and what the possibility of leaking is, so that the system being tested can be assessed regarding its performance before actually putting it into service.
2. *Hydrotesting is not important* because it is not part of the manufacturing process itself. Also, it is cheaper to conduct it with the available water sources than going to rather "fancy" options.

The resultant practice from these two different approaches is that when it's necessary for the job it is done, but it's done in a way that appears to be "cheaper". Nevertheless, it is true that the recommended alternatives to using untreated water are not presented as inexpensive options. Table 5.1 presents some such alternatives.

In these instances, a simple balance between what you get (mitigation programs) and what you lose (corrosion) could be instructive. Javaherdashti in his study of the economy of the treatment of MIC induced by hydrotesting emphasises some important aspects of applying a successful mitigation program [26]. A brief of this study is given below.

Table 5.1 Pros and cons of some alternatives to untreated water for hydrotesting [27]

Alternative	Pros	Cons
Demineralised water (DW)	• Provided that there is draining and drying of the system at the earliest opportunity after hydrotesting, it is recommended. • Disposal usually is not a problem.	• Costly, difficult task of drying a large process system after testing
High-purity steam condensate (HPSC)	• Provided draining and drying the system at the earliest opportunity after hydrotest, it is recommended. • Depending on the chemicals present in the condensate, disposal may be a problem.	• Costly • Practical problems with finding a chemically clean, large steam system for testing a large process system • More problematic than DW approach

5.5 Some Points Regarding the Feasibility of Mitigation of Hydrotesting-assisted MIC

MIC caused by hydrotesting is certainly a case requiring that mitigation programs be launched. For very clear reasons, although all mitigation programs aim at the reduction and elimination of unwanted corrosion, they differ in both extent and application principles. It must be noted that the approach discussed here can be applied to any case of corrosion, whether MIC or not.

Mainly before and after any mitigation program, some steps must be checked; these steps are "confirmation", "mitigation", "control", and "feedback".

What is meant by "**confirmation**" is to prove that the case has been caused by MIC and not, for instance, stray current. Some of the factors that contribute to confirmation step are:

- Vulnerability of the material to corrosion under the working conditions: it is a well-known fact that certain materials are expected to fail more rapidly under certain working conditions.
- Vulnerability of physical, chemical and, in cases of MIC, biological conditions of the system towards corrosion: for instance, when the design of a piping system is such that it produces "dead corners" where a fluid like water can become stagnant and MIC becomes very likely.

The second step is "**mitigation**". This step contains factors such as physical mitigation, chemical mitigation, improving working conditions, design modification requirements, and corrosion knowledge management (CKM). These items have been addressed and discussed in Chapters 2 and 3 of this book.

The third step is "**control**". When the case is challenged and some mitigation programs are advised, it will be good practice to control and see whether the pro-

Table 5.2 Four steps for a good MIC mitigation program

Steps	Factors to be considered
Confirmation	Vulnerability of material to MIC; physical, chemical, and biological vulnerability of the system and working conditions to MIC, including factors such as pH, temperature, the availability of nutrients, and required chemical species such as carbon and nitrogen; the availability of energy source, low-flow or stagnant water, existence of too many branches, or dead corners (in piping systems); existence of corrosion-enhancing bacteria in the system, finding corrosion by products unique to corrosion-enhancing micro-organisms (such as FeS for SRB), in some cases with high care, pit morphology (pit-within-pit) …
Mitigation	Physical mitigation such as pigging for pipelines, chemical mitigation such as use of inhibitors and/or biocides, improving working conditions such as avoiding water stagnation, adjusting pH so that it will not help micro-organisms to grow and act, design modifications of the system, applying corrosion management …
Control	Comparing corrosion rates before and after applying mitigation, comparing performance of the system before and after, cost-effectiveness of the mitigation program …
Feedback	Documentation of working conditions before and after applying mitigation, documentation of the mitigation method(s) for later use …

posed program will work as expected. When one has achieved how to control the case, it will be very useful if the case history along with all the details of confirmation, mitigation, and control are documented (feedback) so that in the future, it will be easy to refer to and consider the mitigation program for any modifications required. More details of these four steps are shown in Table 5.2.

On the other hand, a very important factor in applying a successful mitigation program is to know how to evaluate whether it was both suitable and cost-effective. In other words, one has to see how beneficial the mitigation program was, both technically and economically [27].

Based on several case studies, Kobrin *et al.* recommended some practices that may be useful to prevent MIC induced by hydrotesting [28]. Some of these practices are as follows:

- Always use the cleanest water available; that is, demineralised water, potable water, steam condensate, and the like.
- Within a maximum of 3 to 5 days after a hydrotest, drain and dry the water. "Make this a requirement on purchase orders, engineering specifications, fabrication procedures and drawings."
- Make fabrication crevices as minimal as possible.
- Specify good-quality welds and avoid heat tint scales (use inert gas back-up procedures, for example). Remove the heat tint scale mechanically (grinding, electropolishing and/or abrasive blasting) and/or chemically (pickling).
- Design horizontal pipelines and heat exchangers to be "self-draining".
- At flanged connections, non-wicking gaskets should be used.

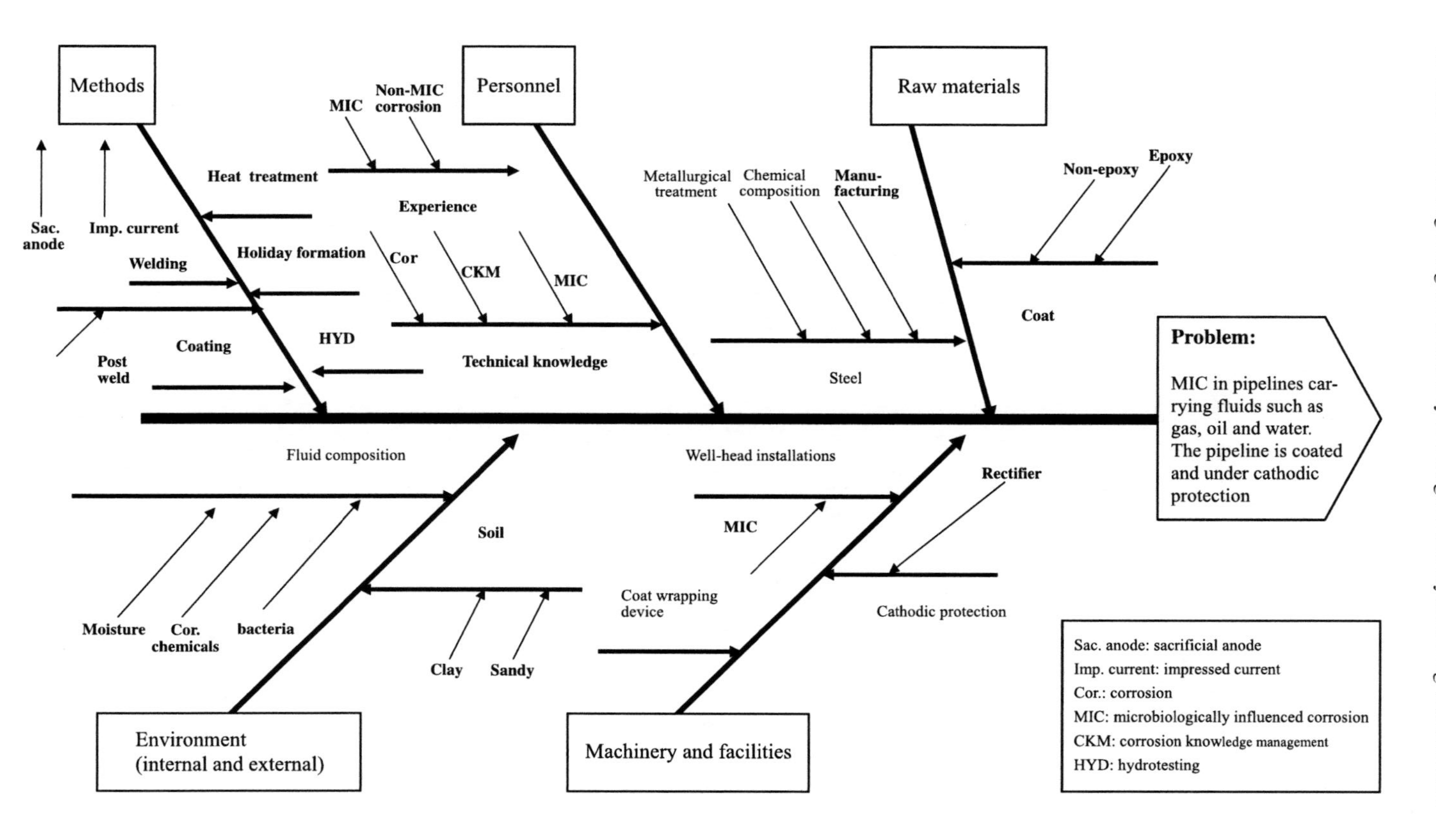

Figure 5.7 Important factors that must be taken into account regarding the possibility of making a system vulnerable to MIC

- While high-flowing water may still develop MIC, avoid designs that promote water stagnation. Design for flow velocities higher than 1.5 m/sec.
- Good material selection and upgrading the existing materials to more corrosion-resistant ones can always be considered.

We add into all these a vibrant, open-minded CKM system specially designed for hydrotesting applications. Figure 5.7 summarises important factors that must be taken into account regarding the possibility of making a system vulnerable to MIC.

5.6 Summary and Conclusions

The answer to the question of "what makes a system vulnerable to MIC?" is certainly a multidimensional one. There are many factors involved in increasing the likelihood of making a system susceptible enough to become deteriorated by microbial corrosion. These factors can be divided into two main categories: (1) physico-mechanical factors such as water velocity or roughness of the surfaces, and (2) chemical factors such as the effects of alloying elements and TDS. Welding and hydrotesting are also very important because if they are not carried out completely and precisely, this can put the system in danger of MIC. What happens in real life is that systems become not only vulnerable to but also contaminated with MIC too soon to be easily detected. For this reason, it is imperative to know how MIC can be detected and what the advantages and disadvantages of these methods are.

The next chapter looks at the methods and techniques that can be applied to detect MIC.

References

1. King RA (1979) Prediction of corrosiveness of seabed sediments. Paper 228, CORROSION/79 March 1979, NACE International, Houston, Texas USA
2. Francis RG, Byrne G, Campbell HS (1999) The corrosion of some stainless steels in a marine mud. Paper 313, CORROSION/99, NACE International, Houston, Texas USA
3. Farinha PA, Javaherdashti R (to be published) Ranking corrosivity of marine sediments on steel structures as induced by sulphate reducing bacteria
4. Farinha PA (1982) Subsediment corrosion of sheet steel piling in ports and harbours with particular refernce to sulphate reducing bacteria. PhD Thesis, University of Manchester
5. Javaherdashti R (2003) Assessment for buried coated metallic pipe lines with cathodic protection: Proposing an algorithm. CORROSION 2003. Pipeline Integrity Symposium. USA
6. Krooneman J, Appeldoorn P, Tropert R (2006) Detection prevention and control of microbial corrosion. Eurocorr Maastricht
7. Torres-Sanchez R, Garcia-Vagas J, Alfonso-Alonso A, Martinez-Gomez L (2001) Corrosion of AISI 304 stainless steel induced by thermophilic sulfate reducing bacteria (SRB) from a geothermal power unit. Materials and Corrosion 52(8):614–618

8. Javaherdashti R (2007) A background fuzzy algorithm for biofilm formation. MIC – An International Perspective Symposium. Extrin Corrosion Consultants, Curtin University, Perth, Australia 14–15 February 2007
9. Scott PJB (2004) Expert consensus on MIC: Failure analysis and control. Part 2. Mater Perform 43(4):46–50
10. Kobrin G (1994) MIC causes stainless steel tube failures despite high water velocity. Mater Perform 33(4):62
11. Al-Hashem A, Carew J, Al-Borno A (2004) Screening test for six dual biocide regimes against planktonic and sessile populations of bacteria. Paper 04748, CORROSION 2004, NACE International, Houston, Texas USA
12. Scott PJB (2004) Expert consensus on MIC: Prevention and monitoring. Part 1. Mater Perform 43(3):50–54
13. Enos DG, Taylor SR (1996) Influence of sulfate-reducing bacteria on Alloy 625 and austenitic stainless steel weldments. CORROSION 52(11):831–842
14. Stott JFD, Skerry BS, King RA (1988) Laboratory evaluation of materials for resistance to anaerobic corrosion caused by sulphate reducing bacteria: Philosophy and practical design: The use of synthetic environments for corrosion testing. ASTM STP 970 Francis PE and Lee TS (eds):98–111 ASTM
15. Scragg, AH (1991) Bioreactors in Biotechnology: A Practical Approach, Chapter 2, Ellis Horwood
16. Walsh D, Pope D, Danford M, Huff T (1993) The Effect of microstructure on microbiologically influenced corrosion. J Matls (JOM) 45:22–30
17. Percival SL, Knapp JS, Wales DS, Edyveab RGJ (2001) Metal and inorganic ion accumulation in biofilms exposed to flowing and stagnant water. Brit Corr J 36(2):105–110
18. Lopes FA, Morin P, Oliveira P, Melo LF (2005) The influence of nickel on the adhesion ability of *Desulfovibrion desulfuricans*. Colloids and Surfaces B: Biointerfaces 46:127–133
19. Kurissery RS, Nandakumar K, Kikuchi Y (2004) Effect of metal microstructure on bacterial attachment: A contributing factor for preferential MIC attack of welds. Paper 04597, CORROSION 2004, NACE International, Houston, Texas USA
20. Duddridge JE, Pritchard AM (1983) Factors affecting the adhesion of bacteria to surfaces. Proceedings of Microbial Corrosion, 8–10 March 1983; The Metals Society; London
21. Borenstein SW (1988) Microbiologically influenced corrosion failures of austenitic stainless steels welds. Mater Perform 27(8):62–66
22. Borenstein SW (1991) Microbiologically influenced corrosion of austenitic stainless steel weldments. Mater Perform 30(1):52–54
23. Brinkley III DW, Moccari AA (2000) MIC causes pipe weld joint problems. Mater Perform 39(6):68–70
24. Javaherdashti R (2003) Enhancing effects of hydrotesting on microbiologically influenced corrosion. Mater Perform 42(5):40–43
25. Iranian Petroleum Standards: Engineering standards for start-up and general commissioning procedures. (1999) IPS-E-PR-280 (0) Section 7.2.7, June 1999
26. Javaherdashti R (2003) A note on the economy of MIC mitigation programs. Proceedings of Corrosion Control and NDT, 23–26 November 2003, Melbourne, Australia
27. Stoecker G (1993) MIC in the chemical industry. In: A practical manual on microbiologically influenced corrosion. Kobrin G (ed), NACE International, Houston, Texas USA
28. Kobrin G, Lamb S, Tuthill AH, Avery RE, Selby KA (1997) Microbiologically influenced corrosion of stainless steels by water used for cooling and hydrostatic testing. Nickel Development Institute (NiDI) Technical Series, No. 10 085. Originally from the paper presented at the 58th Annual International Water Conference, Pittsburgh, Pennsylvania USA, November 3–5 1997

Chapter 6
How Is MIC Detected and Recognised?

6.1 Introduction

In previous chapters, the importance of biocorrosion and its possible mechanisms were discussed. We also looked at some crucial factors that could increase the likelihood of MIC in a given system. The particular focus of the chapter was the avoidance of microbial corrosion. However, almost all the time, what happens in real life is that the system of concern has already been contaminated and the outstanding question is no longer how to prevent, but rather how to estimate the severity and extent of MIC. For instance, while for SRB-induced MIC, some investigators believe that no relationship exists between the corrosion rate and the number of the bacteria cells [1][1], the number of acid-producing bacteria in a system has a profound effect on the corrosion rate.

The distinction between "recognition" and "detection" of MIC has been introduced here to separate those methods that use microbiological means to assess MIC from those that do not. Therefore, our convention here is that the "recognition methods" do not use biology but do use other methods and technologies to deal with the "crime scene investigation" of what the bacteria have done. The "detection methods", on the other hand, are focused mainly on the application and implementation of biological techniques by making use of the features of the bacteria.

[1] In this regard, see also [2]. The general criteria for evaluation of soil microbial corrosivity based on SRB counts alone have been reported by [3] On the other hand, for general criteria for MIC in soil including bacteria such as SRB and iron bacteria, among others, see [4]. The source for both the cited studies regarding a relationship between SRB numbers and the severity of corrosion is the paper by [5]. According to this investigation, if the number of SRB per gram is less than $5*10^3$, there is no risk of MIC, whereas a count of 10^4 or more of SRB per gram of soil is alarming: a severe case of MIC.

6.2 Recognition Methods

6.2.1 Investigating Vulnerable Systems and Components

More often than not, a large percentage of the risk (> 80%) is found to be associated with a small percentage of the equipment item (< 20%) [6]. In other words, there are certain components in a system that could be vulnerable to MIC, and also certain types of materials which are prone to a risk of biocorrosion.

Almost every system that uses untreated liquid water in contact with rough surfaces [7] does have the potential of risking MIC for which time-management is crucial. Currently, there is a relatively good understanding of what particular industries are prone to MIC and where to expect this type of corrosion. Some examples of the components and parts that may suffer most from MIC are open (or closed) cooling systems, water injection lines, storage tanks, residual water treatment systems, and filtration systems, different types of pipes, reverse osmosis membranes, and potable water distribution systems [8]. Also MIC can be expected to occur more often at welds and heat affected zones (HAZ), under deposits and debris and after hydrotesting (inadequate drainage/drying)[2] [9].

Therefore, for example, the possibility of getting MIC in the power generation industry is high, and in this particular industry, cooling systems are more susceptible to MIC than, say, the fire-side of the water walls in the boiler.

Another important issue, of course, is the material itself. A review of case histories shows that, in general, MIC-related failures probably account for less than 10% of total corrosion failures in stainless steel systems [10] (thus putting them well above mild steels). However, in some cases, it is the mild steel that as a material shows good performance with regard to susceptibility to microbial corrosion to MIC: Olesen *et al.* in their investigation of corrosive effects of manganese-oxidising bacteria observed that these bacteria do not cause high rates of corrosion by the deposition of manganese oxides in systems that have been built from mild steel only [11].

Moreover, it must also be noted that different classes of stainless steels do not behave the same when exposed to biological environments. For instance, apparently, grade 304 stainless steel may be regarded as inferior to grade 316 stainless steel from the standpoint of resistance to MIC.

6.2.2 Pit Morphology

In their outstanding review of MIC, Little *et al.* summarised the trend in MIC literature regarding the efforts that have been made to establish a relationship between the shape of the pit (pit morphology) and MIC [12].

[2] Of course here the reader will notice that using untreated water is also an important issue (wrong hydotesting) as mentioned in Chapter 5.

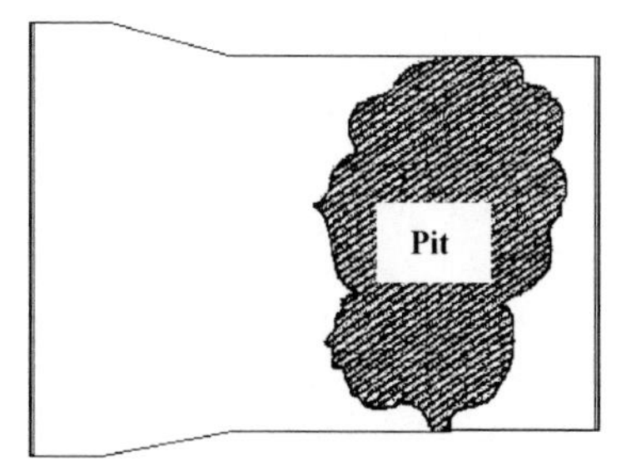

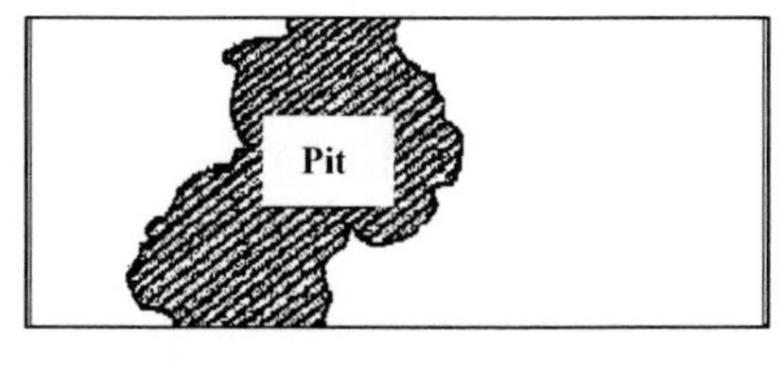

Figure 6.1 Two examples of the so-called "characteristic pit morphologies" associated with MIC in stainless steels

Some researchers and investigators [10, 13] have used or advised the use of "characteristics pit morphologies" in assessments of MIC cases as indicators. Generally, it was believed that a certain pit shape exists that can be applicable to, at least, materials of the same rank. An example is the "gauge"-like pits found in the 300-series of stainless steels (Figure 6.1), where there is a narrow opening and a wide interior space.

Nowadays, it is not believed that such relationships between pit morphology and classification of the attack as microbial corrosion exist. There are reported cases [12] where the diagnoses have been made based only on the pit morphology as an indication of MIC, but have been found to be caused by factors other than MIC. Tatnall and Pope [14] also noted this point by giving the example of corrosivity of ferric chloride on stainless steels or welds and addressing it as "not necessarily ... biologically produced". Stray DC currents, for example, may create corrosion morphologies resembling MIC.

Recent investigations have revealed that the initial stages of pit formation by certain types of bacteria indeed have special characteristics. Some of these cases were reported by Little *et al.* [12]. In a series of investigations on microbial corrosion of pure iron (99.99%) [15] and carbon steel [16] in the presence of SRB *(D. desulfuricans)*, de Romero *et al.* also showed that severe pitting occurred where the bacteria had been formed as colonies. However, in none of these cases has any indication for pit determination by naked eye been given to date.

It seems that with regard to trials aimed at establishing a relationship between the pit morphology and classification of the attack as MIC, the words of wisdom by Tatnall and Pope [14] can be better understood: "surface morphology relates to the chemistry at the metal surface, not to the presence or absence of micro-organisms".

6.2.3 Mineralogical Finger Prints

Another approach that can be used to identify cases of microbial corrosion is by determination of the minerals formed, as some of them do form only under microbial conditions.

The presence of SRB can be often justified by the presence of Fe_xS_y minerals such as troilite (FeS), pyrrhotite ($Fe_{0.875-1S}$), mackinawite ($FeS_{0.93-0.96}$), greigite (Fe_3S_4), or amorphus iron sulphide (FeS_{amorph}) [17, 18]. It has been reported that the primary films formed by SRB on steel are mackinawite and a protective film of siderite ($FeCO_3$) [19].

Both then are transformed to cubic greigite and greigite with a rhombohedral structure, *i.e.*, smythite. Eventually, the final stage of all these transformations is the formation of non-protective pyrrhotite under anaerobic conditions. Pyrrhotite may form after nine months [12]. For copper-nickel alloys affected by SRB, the fingerprint minerals could be djurleite, spinonkopite, and high-temperature polymorph pf chalcocite [12]. Figures 6.2a and b show iron sulphide film formed on two types of mild steel coupons exposed to SRB culture. Figure 6.2a [20] shows the iron sulphide

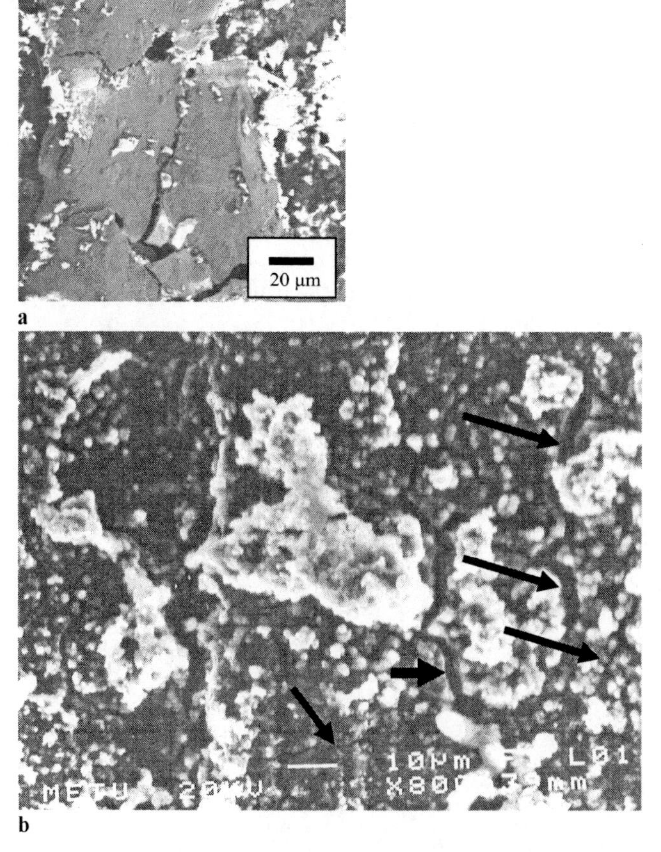

Figure 6.2 **a** Scanning electron microscope (SEM) micrograph of the cracked corrosion product film (most probably FeS) that had been formed on mild steel electrode exposed to pure SRB environment after the biofilm removal **b** SEM of another type of carbon steel (N-80) after being exposed to SRB culture. Some of the cracks are shown by arrows in both pictures

film formed on carbon steel with a composition (wt%) of carbon (0.25%) and chromium (0.3%). The composition (wt%) of the carbon steel shown in Figure 6.2b [21] is also carbon (0.39%) and chromium (0.11%). As seen in the figures, both steels show cracks, indicating that the films formed are not mechanically stable enough to resist external forces such as shear forces caused by intermittent flow patterns. We should mention an important point here: insufficient treatment of MIC is, in most instances, worse than no treatment at all. An example is an increasing flow of water or inducing oxygen after a period of stagnation without having a versatile mitigation program. In both "treatments" mentioned above, if necessary precautions are not taken, the existing situation may become worse: for instance, by increasing the flow, the shear forces thus produced are capable of damaging the non-protective, brittle films and therefore producing cracks and enhancing corrosion.

When a crack forms, the local oxygen concentration (pressure) within the crack and in the area around it will be different. Due to the lack of access to the inside of the crack, the oxygen partial pressure within the crack decreases whereas the adjacent area is still rich in oxygen. Furthermore, these cracks can establish electrochemical cells such as differential aeration cells where the low partial pressure of oxygen in one side makes it anode and, as a consequence, the likelihood of corrosion also increases.

6.2.4 Appearance and Colour of Corrosion Products

The colour of the corrosion products can also provide good clues in an investigation of MIC and, probably, the likely species involved.

Black colour, smelly iron sulphide corrosion products, and reddish-brown colour deposits may be good indicators of SRB and iron-oxidising bacteria, respectively [22]. Figures 6.3a and b show a piece of carbon steel before and after it was exposed to an SRB culture. On the other hand, when iron-reducing bacteria are present and actively reducing iron, the dark greenish colour is a good indication of the presence of these bacteria (Figure 6.4).

In the case of sulphur-oxidising bacteria (SOB), the colour of the corrosion products is reportedly yellow [23].

6.2.5 Characteristics of Microbial Attack

There are a few patterns to address the microbial attack that can be used to evaluate a case of MIC, as very briefly addressed and described below [24]:

1. Attack by physical presence and attachment of microbial cells onto the surfaces (as it occurs in the electronic industry by adhering cells onto the electronic chips, thus requiring having very clean air in the laboratories)

a

b

Figure 6.3 Appearance of mild steel **a** before and **b** after about two months of exposure to SRB culture [20]

Figure 6.4 Cultures of active IRB, the two test tubes on the left show positive ferric iron reducing culture in different cell concentrations, the last tube on the right shows negative ferric iron reducing culture. (Courtesy of Extrin Consultants)

2. Attack by the excretion of mineral or organic acids of a biological source such as hydrated hydrogen sulphide gas (by SRB, for example) or nitric acid (by, for instance, nitrifying bacteria) that results in hydrolysis of the material
3. Attack by organic solvents that are produced as a result of the actions of fermentative bacteria
4. Attack caused by the salt which itself may have been produced as the result of reactions between anions (which are final products of microbial metabolism) and cationic components of ceramic materials. The results of these reactions could range from swelling of the porous material (due to the hydration of these

often highly water-soluble salts) to blasting (caused by dryness and thus formation of voluminous crystals) to freeze-thaw attack (the physical attack that may be the result of the swelling originally resulting from microbial activity)
5. Attack by the effect of biofilms that may cause, among others, problems such as localised corrosion and reduced flow velocity
6. Attack by enzymes excreted by micro-organisms that live on insoluble compounds to turn them into smaller fragments. An example is the deterioration of cellulose (wood) into glucose. This matters if the insoluble compounds are impregnated by organic substances (such as resins or waxes) used to improve their features and characteristics. Therefore, degradation of such organic additives by certain micro-organisms may cause serious problems for the performance of the material.
7. Attack which is stimulated by the solubility action of most organic acids, capable of complexing metal ions which may be otherwise insoluble/low-soluble products. Also, emulsifying agents produced by micro-organisms (*e.g.*, phospholipids) can degrade "insoluble" materials such as pyrite or low-soluble items such as elemental sulphur.

There are some additional pertinent points, summarised as follows [23, 25]:

- If osmium compounds such as osmium tetroxide are used for fixation of the sample chemically, the osmium picked by the micro-organism can make it easier to distinguish samples with biological cells from inorganic debris and crystals of approximately the same dimensions and shape.
- If by using methods such as EDXA or atomic absorption spectroscopy, total or organic carbon is above 20% and the sulphur is about 1% or more (in the absence of any other source of sulphur), or wet chemical methods show that there are high concentrations of chlorides in fresh water and iron or manganese in non-ferrous materials, or phosphorus in corrosion products (unless coming from treatments such as phosphate treating) and very low nickel (below the material ratio), all these clues can signal the possibility of a microbial attack.
- There could be other methods such as the use of sulphur isotopes that have a lighter atomic number so that when they react with micro-organism-driven sulphate reduction (for example SRB), they can easily be identified from the non-biologically formed heavier sulphur isotope compounds.

There is still no united idea regarding the so-called biogenic iron sulphide as a good indicator of bacterial activity compared to abiotic sulphide films (please also refer to footnote 20 in Chapter 4 regarding Smith and Miller's review). Some researchers have postulated that the presence of mackinawite may be used as an indication of MIC by SRB [26]. On the other hand, others believe that while SRB may affect the crystallisation mode of iron sulphides, they reject the idea that mackinawite is the "unique" SRB-influenced corrosion product [27]. Thus, until researchers agree on one or another idea, it is not recommended to use sulphide films as the only way to identify MIC.

6.3 Detection Methods

There are some methods and techniques used by microbiologists to assess MIC cases. All these methods have their pros and cons. We will be dealing with each very briefly here without going through the microbiological details of each; more details can be found in the given references.

6.3.1 Culturing and Its Alternatives

By "culturing", what is basically meant is that the nutrients and the temperature necessary for the growth of certain types of bacteria are prepared and in the course of time, the bacterial species of interest are grown. Even with this definition, one can easily see the main drawback of culturing methods: you get what you have asked for. In other words, if the environment (in terms of chemicals used as nutrients, oxygen, and the temperature) are within the range of a prearranged band, this means that the bacterial species that do not meet those criteria will not grow. That is why in a culture prepared for room-temperature (mesophilic), neutral pH-tolerant, strictly anaerobic bacteria such as some species of strictly anaerobic SRB, one cannot grow, say, aerobic (oxygen-demanding), acidophil (acid-loving), thermophilic (high-temperature) sulphur-oxidising bacteria. In addition to this, culturing can reveal only 1% or less of the total bacteria present in a given sample [28]. Therefore culturing methods may not be regarded as very reliable methods on their own, as they may produce dubious results which are open to discussion.

Maxwell *et al.* oppose such shortcomings of culturing methods by stating that the very low percentage of culturable species "is not unique" to a certain industry such as the oil industry [29]. They also emphasise the importance of many issues such as (a) the possible positive impacts of more investment by industry to enable the laboratories to use a wider range of culture media, (b) increasing the reliance and qualification of the testing methods by applying statistically relevant techniques (such as triplicate Most Probable Number method), and also (c) the "fact" that "in the hands of trained microbiologists, similar techniques [such as Culturing and Extinction serial dilution] are employed as useful tools in clinical, pharmaceutical and other industrial situations".

While it is very true that without more funds, progress in the field of MIC – like other disciplines – is almost impossible, comparing culturing method outcomes in medical applications with industrial (engineering) applications is oversimplifying the situation. For one thing, where a disease and its cause are concerned, there is not a wide range of micro-organisms that can be considered as being related. Every time a similar pathological situation arises, it is that particular type of micro-organism which is responsible. However in the case of, for example, accelerated low water corrosion and buried pipelines (to be studied in more detail in Chapter 7), certainly more than one type of bacteria can be involved. Therefore, by just relying on the culturing method, one can miss a lot of such bacteria. Another interesting example is when a slime sample from a cooling water system is diagnosed as having high numbers of *Pseudomonas.* Practically, such diagnoses may not be that important as "rather than being a single strain, *Pseudomonas* is a diverse genus of common, ... aerobic organisms associated with everything from industrial biofouling to urinary tract infections" [14]. Therefore, the complexity of MIC in industrial systems used for engineering purposes makes culturing method easy and inexpensive yet least reliable as a complete method of detection.

Little *et al.* reported a very interesting case where the corrosion rate of the mild steel specimen exposed to a mixed culture of SRB and oil-oxidising bacteria changed according to the medium used [30], demonstrating how the culture medium composition can affect the corrosion rate and intensity. Setareh and Javaherdashti compared the precision of some SRB detection cultures and found out that by modification or using new formulations of relatively well-known culture media such as API RP38 and Butlin, the resultant synthetic cultures could perform relatively better in terms of getting more positive results [31]. These results confirm again that there is no such thing as "the best standard culture", as long as you know what you want to find. Table 6.1 summarises pros and cons of some of common detection methods that are used for dealing with bacterial samples suspected of containing corrosion-related bacteria.

No discussion of the detection methods applied in MIC studies is complete without mentioning molecular biology methods, which are becoming more and more popular in this area. We will briefly discuss these methods here, without going through all the details.

Due to the shortcomings of traditional, relatively inexpensive techniques like culture-dependent methods,[3] other methodologies have been developed. These methods are based on molecular biology techniques and use the genetic material (DNA and RNA) of the bacteria. Referring to them collectively as "genetic techniques", they have the capacity to [32]:

- highlight the dominant bacteria in a natural sample (an ecological system) that contains other species as well with no problems regarding the serious limits of common viable counting methods
- determine the relative proportion of MIC-related bacteria within the whole bacterial community
- identity the bacteria which are resistant to biocides

---

[3] One of the culture-dependent methods is the Most Probable Number (MPN) that has been reported to "underestimate the size and misrepresent the composition of microbial communities"; see [35].

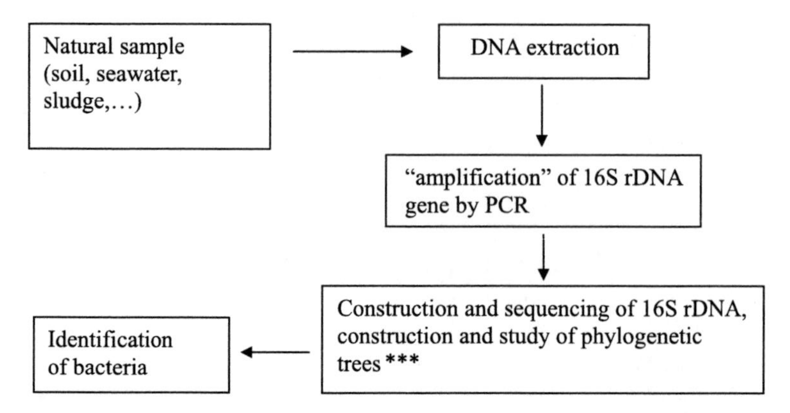

Figure 6.5 Basic steps involved in genetic techniques to be used for MIC studies

- evaluate the population size changes imposed by using biocide or nutrient modifications
- be counted as a more reliable method for sampling biological samples which is not affected by time or transportation factors[4]

Genetic techniques mainly analyse the DNA extracted from cultures or environmental samples.[5] The extracted DNA will be used to find "molecular fingerprints" using techniques such as polymerase chain reaction (PCR). PCR is used to "amplify" *in vitro* the DNA fragments that then can be used to identify the bacteria [32]. Figure 6.5 schematically presents the main steps involved in genetic techniques using PCR.

To make the traditional PCR methods, new techniques such as real-time quantitative PCR (qPCR) have been developed and used, whose advantages over traditional PCR have been reported to be: (a) providing more accurate and reproductive quantitative data regarding microbial communities, (b) having a detection range of six orders of magnitude or more, (c) no need for post-PCR manipulation and treatment, and (d) much better analysis capability [34].

---

[4] Sampling and preserving the samples could be a really tough issue, especially due to the practicality of these practices. If sampling and handling are carried out in a manner in which the microbial samples get contaminated or die during transportation, the results could become highly dubious.

[5] "Phylogenetic trees are graphical representations of the evolutionary relatedness of a group of organisms", in [33].

Table 6.1 Pros and cons of some common detection methods used in MIC studies [14, 22, 35]

Detection Method	Some advantages	Some disadvantages
Culturing	Rather simple and routine, comparatively unsophisticated and cost effective	Only a very small portion of the whole micro-organisms are cultivable; it may be too time-consuming; the culturing environment can be different from the natural environment of the micro-organism; it is selective, *i.e.*, it allows growing of "pre-selected" micro-organisms
Direct microscopic Examination	A good way to estimate total number of bacteria by using staining/counting techniques; It can be used for direct inspection of certain large, distinctively shaped micro-organisms (such as filamentous iron bacteria and stalked iron oxidisers such as *Gallionella*)	This method cannot identify a large number of bacteria, so it must not be used alone and requires involvement of other detection methods as well.
Adenosine triphosphate (ATP) assay	This method has been used to estimate relative total bacteria number in environments where "non-bacteria" ATP is rare (*e.g.*, oil field water samples); It can be carried out within much shorter times (less than an hour)	This method cannot distinguish between ATP extracted from bacteria and other organic debris in the sample; the amount of ATP is not predictable in SRB and some other common environmental bacteria, giving rise to very rough approximate values for total bacteria count.
Anti-body tests	Commercially available only for SRB; it is rapid (between 20 min. and an hour) and inexpensive (provided that there is microscopy available)	The sensivity limit is normally approx. 10^3 cells/mL; however, its "total number of bacteria" counts must be checked with culturing methods as well; cannot distinguish between living and dead bacteria.
Fatty acid analyses	Commercially available; its precision is better than anti-body techniques in "marking" available micro-organism types.	Relatively expensive; less quantitative; limited library of fatty-acid signatures of bacteria; cannot detect certain bacteria such as iron oxidisers.

6.3.2 *Quick (or Rapid) Check Tests*

Most of the time, what is required is to determine whether or not MIC and particularly SRB are part of the problem. In these conditions, one needs a prompt answer; although it may not be very accurate, at least it will give an idea about the "presence" or "absence" of SRB. However, one must be very cautious when using this terminology as it may also be misleading, as it will be discussed below.

These methods can be classified as "absence/presence tests", in terms of microbiological tests. The tests of this kind that are most frequently used in the field are an "acid test" and a "lead acetate test". In an acid test, a few droplets of diluted hydrochloric acid are added on the corrosion products. If there is a smell of "rotten egg" – characteristic of hydrogen sulphide gas – then the corrosion products do contain sulphides. The "lead acetate test" is basically using moist lead acetate $[Pb(C_2H_3O_2)_2 \cdot 3H_2O]$ papers. In contact with sulphides, these papers turn black.

While these test methods are easy and straightforward, one should bear in mind that they are only useful when it has been established with confidence that the case is indeed MIC and does involve SRB. The reason is that by conducting these tests, what is actually tested is the presence of sulphide, not SRB. In other words, if the case is confirmed as MIC and it is also confirmed that SRB are involved, then these tests can be called "absence/presence tests" in their microbiological context, meaning that they can be used to provide proof of the existence of any microorganisms that are capable of producing sulphides. Otherwise they just show the presence of sulphides, and these sulphides may have been created by non-biological means. These tests, at best, are capable of confirming the presence of SRB. They cannot be used to test the presence of manganese oxidising bacteria. So, if you have a mixed sample (more accurately, a mixed culture) that contains many types of bacteria, using the acid test and lead acetate test may not be useful.

As a result, bearing in mind the limitations of both of these rapid test methods, they may become very handy to help build a story for a corrosion case where there is no immediate access to laboratory facilities.

Therefore, if the detection test is not sophisticated, in terms of both the testing techniques and the means, it can be done on site and requires no background science of either microbiology or corrosion, we call these tests "rapid check tests". In this regard, a majority of the tests called "rapid check tests" by some experts [14, 36] will need to be re-assigned. To give an idea of what these so-called quick check tests are, here is a list of the test types:

1. Microscopic direct method evaluation
2. ATP luminescence
3. Hydrogenase test
4. Desulphovirdin test
5. APS-reductase test
6. Enzyme linked immunosorbent assay (ELISA)
7. Auto-radiography
8. Fatty acid analyses

Both the APS-reductase and hydrogenase tests work on the basis of tracing enzymes: one tracks down APS-reductase and the other tracks down hydrogenase. However, it must be noted that while the hydrogenase test can only be done on SRB that have this enzyme only, APS-reductase-based tests are applicable to all SRB. The main drawback is, of course, it can only detect SRB. Besides, although the commercial kits are user-friendly, a certain level of hands-on experience is needed for carrying out the tests more confidently.

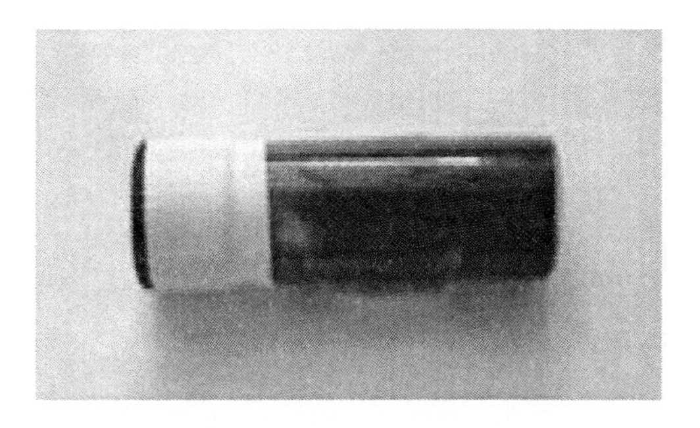

Figure 6.6 SRB-positivity, as evidenced by a blackened Sanicheck® SRB kit (courtesy of Extrin Consultants)

Another point is that some of the commercially available SRB-detection quick tests normally rely on visual effects caused by SRB presence. One of these visual factors is the blackening, as shown in Figure 6.6.

6.3.3 Electrochemical Recognition Methods and Their Application to MIC

Microbial corrosion is a multi-disciplinary topic, and thus many disciplines must be involved for the assessment of its effects. Electrochemistry is definitely one of these disciplines. When electrochemical experimental techniques are applied to non-living objects, there is no fear of altering/modifying the environment. However, when it comes to living micro-organisms, it becomes another story.

The examples below suggest how applying a voltage on microbial communities can have adverse effects:

- A report on cathodic protection effects of steel pipes against MIC [37] suggests that, under laboratory conditions, applying voltages with values more negative than $-0.95\ V_{Cu\text{-}CuSO_4}$ may decrease the number of iron bacteria as a result of environmental changes caused by cathodic protection process.[6]
- Again with regard to cathodic protection criteria (applying voltages more negative than $-0.95V_{Cu\text{-}CuSO_4}$), it has been reported that applying voltages up to $-1.1V_{Cu\text{-}CuSO_4}$ not only failed to prevent the growth of bacteria on the metal surfaces, it rather prompted the growth of certain microbial species and rate of corrosion [39].
- Little *et al.* [12] reported that in one of their investigations, applying electrochemical polarisation could influence the number and types of bacteria associated with the surface.

[6] Although in this report the type of the bacteria (IOB or IRB) was not specified, from general recognition of iron bacteria (see [38]), it may be anticipated that it was iron-oxidising bacteria whose number had been adversely affected by applying voltage.

The following is a list of some pros and cons of traditional electrochemical methods when they are applied to biocorrosion studies) [40–43]:

Open-circuit potential (corrosion potential), or briefly, OCP:

- By this method, the corrosion potential of a corroding metal is measured by determining the voltage difference between the metal immersed in a corrosive medium and a suitable reference electrode, which is usually a saturated calomel electrode.
- Advantage: Because of its simplicity, the measurement of the corrosion potential has been used in MIC studies for many years. It can be used both in the laboratory and in the field.
- Disadvantage: It measures both anodic and cathodic processes simultaneously and only assesses trends.

Reduction-oxidation (Redox) potential:

- It is pertinent to the relative potential of an electrochemical reaction under no net flow of electrical current (equilibrium conditions). The redox potential in general is a measure of the oxidising power of the environment.
- Advantage: It can be used both in the lab and in the field.
- Disadvantage: It is not useful for evaluating corrosion rates. It requires the simultaneous measurement of the medium pH, because it may result in difficulties in both taking accurate measurements and interpreting the obtained data. Such cases happen when immersion times are not chosen carefully, so that microbial colonisation of the measuring electrode occurs and the measured value will correspond to the chemistry at the electrode under the biofilm rather than to that of the bulk environment. Also, the redox potential measurements of electrochemical reactions must be made under equilibrium conditions where it is usually unlikely to be encountered in real-life experiences performed on living systems such as microbial communities.

Tafel polarisation:

- In this method, applied potential to the system is plotted *vs.* the logarithm of the current density. The resulting curves would intersect at a point representing the corrosion potential and the corrosion current density. In the vicinity of the corrosion potential, the measured log current *vs.* potential curves both deviate from linearity. Nevertheless, both often contain linear segments referred to as Tafel regions.
- Advantage: It can be used in the laboratory or in the field because of the easy interpretation of data.
- Disadvantage: The measurement of corrosion current is dependent on both a steady corrosion potential and the ability to identify the linear Tafel region. Electrolytes in which more than one reduction reaction take place or in which concentration polarisation occurs exhibit less distinct linear regions. Large polarisations may change the electrochemical conditions at the metal surface and could be deleterious to micro-organisms in the biofilm. For systems like some

stainless steels in seawater, in which corrosion potential drifts or fluctuates with time, Tafel polarisation is practically meaningless.

Potentiodynamic sweep techniques:

- The applied potential is increased *vs.* log current and plotted. For a given corroding metal, the corrosion potential and corrosion current will be determined by the point at which the cathodic curve intersects the anodic curve. One of the main experimental variables that can be manipulated is the sweep rate. High scan rates (about 60 V/h) are used to show regions where intense anodic activity is likely. Slower scan rates (1 V/h) are used to identify regions in which relative inactivity is likely, such as stable metal surface conditions.
- Advantage: Useful to predict the corrosion behaviour of passive metals in biotic media containing biofilms. Quantification of microbial effects and rapid scan rates for film-free metals are possible.
- Disadvantage: Results depend on the sweep rate and experimental conditions. Slow sweep rates can affect localised conditions at the metal/solution interface.

Polarisation resistance method:

- The method is based on the linear relationship between changes in the applied potential and the resulting current density when the applied potentials are within ±10 mV of the corrosion potential. The slope of the potential/current curve is approximately linear and has the units of resistance.
- Advantage: Rapid and easy interpretation of the results; it shows a good correlation with the weight-loss method.
- Disadvantage: It is not useful to assess localised corrosion. The presence of biofilms complicates the linear polarisation interpretation by introducing additional electrochemical reactions, which can lead to non-linear polarisation behaviour.

Generally, all direct current polarisation methods, that is, Tafel polarisation, potentiodynamic sweep techniques, and the polarisation resistance method apply voltage to the test environment, that in the case of MIC studies is the microbial environment. The net result of this is altering the environment in such a way that is likely to affect the micro-organisms. Some examples of such adverse effects will be addressed later in this text. It appears that the method showing less adverse influence to the microbial environment is the open-circuit potential measurement. Nowadays, the general trend among investigators seems to prefer OCP and electrochemical noise potential (EPN) as the safest electrochemical recognition methods, as they do not impose voltages upon microbial communities [11, 42, 44, 45].

6.4 Summary and Conclusions

The first step for the recognition and detection of MIC is to try to prove that the case is not MIC-related at all! By doing so, many prejudices and problems associ-

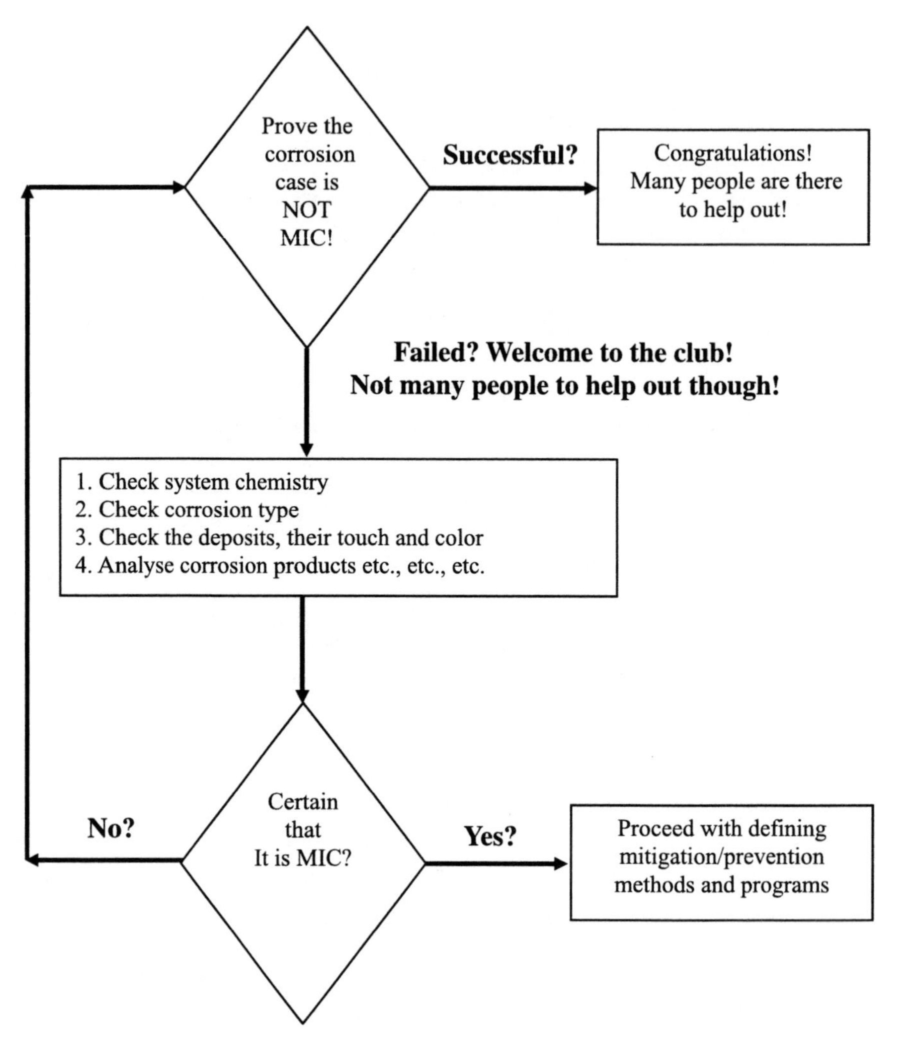

Figure 6.7 A guide for the confirmation steps that may need to be taken to prove that the corrosion case can "really" be MIC

ated with it can be solved.[7] Figure 6.7 shows the required chain of action in the form of a flow chart of the "do's and don'ts" that a corrosion engineer should

[7] Just imagine the situation which is not MIC-related, but due to the insistence of the engineer, the management spends heaps of money and then it is realised that the case was not an example of microbial corrosion at all. It must be the hardest imaginable task to convince the same management about another case that indeed could be microbially induced corrosion. All this could have been prevented if the engineer in charge had first investigated the possibility of non-MIC corrosion. Another extreme is, of course, denying MIC all together; Tatnall describes such misinterpretations as addressing cases where microbial tuberculation corrosion of steels was called

consider in the first place. After you are convinced that the case is microbial in nature, the microbial detection and non-microbial recognition techniques and methods can be applied. This chapter explained some of these methods with a brief summary of their pros and cons.

The next chapter takes the examples of some industrial systems and looks at some common characteristics that can result in MIC although the systems are not technically or industrially similar to each other.

References

1. Little BJ, Wagner P (1997) Myths Related to Microbiologically Influenced Corrosion, Materials Performance (MP), vol. 36, no. 6, pp. 40–44
2. Ilhan-sungur E, Cansever N, Cotuk A (2007) Microbial Corrosion of Galvanized Steel by a Freshwater Strain of Sulphate Reducing Bacteria (*Desulfovibrio* sp.), Corrosion Science, vol. 49, no. 3, pp. 1097–1109, March 2007
3. Mizia RE, Alder Flitton MK, Bishop CW, Torres LL, Rogers RD, Wilkins SC (2000) Long Term Corrosion/Degradation Test, First Year Results, Idaho National Engineering and Environmental Laboratory
4. Stein AA, (1993) MIC Treatment and Prevention in A Practical Manual on Microbiologically-influenced Corrosion, Kobrin G, (ed.), NACE, Houston, TX, USA
5. Ronay D, Fesus I, Wolkober A (1987) New Aspects in Research in Biocorrosion of Underground Structures, Corrosion' 87, Brighton, UK
6. Hovarth RJ (1998) The role of the corrosion engineer in the development and application of Risk-based inspection for plant equipment Mater Perform 37(7):70–75
7. Javaherdashti R (2007) How to deal with MIC? Tips for industry. MIC – An International Perspective Symposium; Extrin Corrosion Consultants, Curtin University, Perth, Australia, 14–15 February 2007
8. Al-Darbi MM, Agha K, Islam MR (2005) Modeling and simulation of the pitting microbiologically influenced corrosion in different industrial systems. Paper 05505, CORROSION 2005, NACE International, Houston, Texas USA
9. Scott PJB (2004) Expert consensus on MIC: Prevention and monitoring. Part 1. Mater Perform 43(3):50–54
10. Jack TR (2002) Biological corrosion failures. Published by ASM International
11. Olesen BH, Nielsen PH, Lewandowski Z (2000) Effect of biomineralized managanese on the corrosion behaviour of C1008 mild steel CORROSION 56(1):80–89
12. Little BJ, Lee JS, Ray RI (2006) Diagnosing microbiologically influenced corrosion: A state of the art review. CORROSION 62(11):1006–1017
13. Cubicciotti D, Licina GL (1990) Electrochemical aspects of microbially induced corrosion. Mater Perform 29(1):72–75
14. Tatnall RE, Pope DH (1993) Identification of MIC. Ch. 8, In: a practical manual on microbiologically influenced corrosion Kobrin G (ed), NACE, Houston, Texas USA
15. de Romero M, Urdaneta S, Barrientos M, Romero G (2004) Correlation between *Desulfovibrio* sessile growth and OCP hydrogen permeation corrosion products and morphological attack on iron. Paper No. 04576, CORROSION 2004, NACE International, Houston, Texas USA

---

water corrosion or under-deposit corrosion by "those [corrosion engineers] who do not understand (or believe in) the biological factors" [46].

16. de Romero M, Duque Z, Rodriguez L, de Rincon O, Perez O, Aranjo I (2005) A study of microbiologically induced corrosion by sulfate-reducing bacteria on carbon steel using hydrogen permeation. CORROSION 61(1):68–7 *et al* 5
17. Lee AK, Buehler MG, Newman DK (2006) Influence of a dual-species biofilm on the corrosion of mild steel Corrosion Science 48(1):165–178
18. Liu H, Xu L, Zeng J (2000) Role of corrosion products in biofilms in microbiologically induced corrosion of carbon steel Brit Corr J 35(2):131–135
19. Tiller AK (1983) Microbial corrosion Proceedings of Microbial Corrosion, 8–10 March 1983, The Metals Society, London
20. Javaherdashti R (2005) Microbiologically influenced corrosion and cracking of mild and stainless steels. PhD Thesis, Monash University, Australia
21. Javaherdashti R, Sarioglu F, Aksoz N (1997) Corrosion of drilling pipe steel in an environment containing sulphate-reducing bacteria. Intl J Pres Ves Piping (73):127–131
22. Blackburn FE (2004) Non-BIOASSAY techniques for monitoring MIC. Paper 04580. CORROSION 2004, NACE International, Houston, Texas USA
23. Scott PJB (2004) Expert consensus on MIC: Failure analysis and control. Part 2. Mater Perform 43(4):46–50
24. Sand W (1997) Microbial mechanisms of deterioration of inorganic substrates – A general mechanistic overview. Intl Biodeterioration & Biodegradation 40(2–4):183–190
25. Scott PJB (2000) Microbiologically influenced corrosion monitoring: Real world failures and how to avoid them. Mater Perform 39(1):54–59
26. McNeil MB, Little BJ (1990) Technical Note: Mackinawite formation during microbial corrosion. CORROSION 46(7):599–600
27. Newman RC, Rumash K, Webster BJ (1992) The effect of pre-corrosion on the corrosion rate of steel in natural solutions containing sulphide: Relevance to microbially influenced corrosion. Corrosion Science 33(12):1877–1884
28. Yee GG, Whitbeck MR (2004) A microbiologically influenced corrosion study in fire protection systems. Paper No. 04602. CORROSION 2004, NACE International, Houston, Texas USA
29. Maxwell S, Devine C, Rooney F, Spark I (2004) Monitoring and control of bacterial biofilms in oilfield water handling systems. Paper 04752. CORROSION 2004 NACE International Houston Texas USA
30. Little B, Lee J, Ray R (2007) New development in mitigation of microbiologically influenced corrosion MIC – An International Perspective Symposium; Extrin Corrosion Consultants, Curtin University, Perth, Australia 14–15 February 2007
31. Setareh M, Javaherdashti R (2003) Precision comparison of some SRB detection methods in industrial systems. Mater Perform 42(5):60–63
32. Videla H (2007) Biofilms in pipelines and their treatment in the oil industry. MIC – An International Perspective Symposium; Extrin Corrosion Consultants, Curtin University, Perth, Australia, 14–15 February 2007
33. Devereux R, Stahl DA (1993) Phylogeny of Sulfate-Reducing Bacteria and a Perspective for Analayzing Natural Communities, in The Sulfate-Reducing Bacteria: Contemporary Perspectives, (Odom JM, Singleton Jr. R, eds.), Springer-Verlag, New York
34. Zhu XY, Modi H, Ayala A, Kilbane JJ (2006) Rapid detection and quantification of microbes related to microbiologically influenced corrosion using quantitative polymerase chain reaction CORROSION 62(11):950–955
35. Le Borgne S, Jan J, Romero JM, Amaga M (2002) Impact of molecular biology techniques on the detection and characterization of micro-organisms and biofilms involved in MIC. Paper No. 02461. CORROSION 2002, NACE International, Houston, Texas USA
36. King RA (2007) Trends and developments in microbiologically induced corrosion in the oil and gas industry. MIC – An International Perspective Symposium; Extrin Corrosion Consultants, Curtin University, Perth, Australia, 14–15 February 2007
37. Kajiyama F, Okamura K (1999) Evaluating cathodic protection reliability on steel pipes in microbially active soils. CORROSION 55(1):74–80

38. Standard Test Method for Iron Bacteria in Water & Water-formed Deposits, ASTM D932-85 (Re-approved 1997), ASTM annual book, ASTM, USA, 1997
39. Pope DH, Zintel TP, Aldrich H, Duquette D (1990) Efficacy of biocides and corrosion inhibition in the control of microbiologically influenced corrosion. Mater Perform 29(12):49–55
40. Videla HA (1996) Manual of Biocorrosion. CRC Press Inc.
41. Stott JFD, Skerry BS, King RA (1988) Laboratory evaluation of materials for resistance to anaerobic corrosion caused by sulphate reducing bacteria: Philosophy and practical design. The use of synthetic environments for corrosion testing. ASTM STP 970 Francis PE and Lee TS (eds) 98–111 ASTM
42. Dexter SC, Duquette DJ, Siebert OW, Videla HA (1991) Use and limitations of electrochemical techniques for investigating microbial corrosion CORROSION 47(4):308–318
43. Dexter SC (1995) Microbiological effects in corrosion tests and standards: Application and interpretation. Baboian R (ed) ASTM Manual Series: MNL 20, ASTM
44. Jack TR, Ringelberg DB, White DC (1992) Differential corrosion rates of carbon steel by combinations of Bacillus sp *Hafnia alvei* and *Desulfovibrio gigas* established by phospholipid analysis of electrode biofilms Corrosion Science 33(12):1843–1853
45. Michael JF, White DC, Isaacs HS (1991) Pitting corrosion by bacteria on carbon steel determined by the scanning vibrating electrode technique. Corrosion Science 32(9):945–952
46. Tatnall RE, (1991) Case Histories: Biocorrosion, in Biofouling and Biocorrosion in Industrial Water Systems, Flemming HC, Geesey GG (eds.), Springer-Verlag Berlin, Heidelberg, Germany

Chapter 7
Examples of Some Systems Vulnerable to MIC

7.1 Introduction

It is not a rare accident to meet people who, despite having no blood connection, look so similar to each other.

No matter how different such individuals may be in other details of their lives, the most interesting features are that they look so much like each other. These "similar, yet, different" characteristics can also be seen in many industrial systems and their problems, especially if MIC is the problem.

As we shall see, the proposed cyclic mechanisms of MIC are very similar in a buried pipeline to accelerated low water corrosion of steel piles of a jetty or wharf. Although there are many aspects of biocorrosion not yet clear, some "rules of thumb" can still be developed to allow estimating the vulnerability of a system

How two different individuals may look alike! (source: www.marshal-modern.org)

to MIC, as stated in detail in Chapter 5. Despite the limitations and related uncertainties, it is still possible to come up with some patterns that repeat themselves in systems where corrosion is enhanced by microbial corrosion mechanisms. It is these general patterns and global features that we are trying to address in this chapter for industrial systems as diverse as firewater lines, off-shore platforms, buried metallic pipelines, and immersed piles.

7.2 Buried Metallic Pipelines

According to the principles of CKM as discussed in Chapter 3, the first step in understanding corrosion is to be able to define the system in which one is interested to detect, define, and mitigate corrosion (or more specifically as the topic of this book is concerned, microbial corrosion). As Figure 7.1 suggests, the following corrosion systems can be defined in a buried pipe:

1. External corrosion system that includes corrosion problems such as those occurring in the soil surrounding the buried pipe, the coating, the cathodic protection system, ...
2. Internal corrosion system, including corrosion problems that are likely to occur with regard to the fluid (water, gas, oil, its temperature and pH, its velocity, its TDS, ...), the lining, ...
3. The corrosion system itself, the pipe, where corrosion can be the result of wrong/incomplete hydrotesting, the steel characteristics (physical, chemical, metallurgical), ...

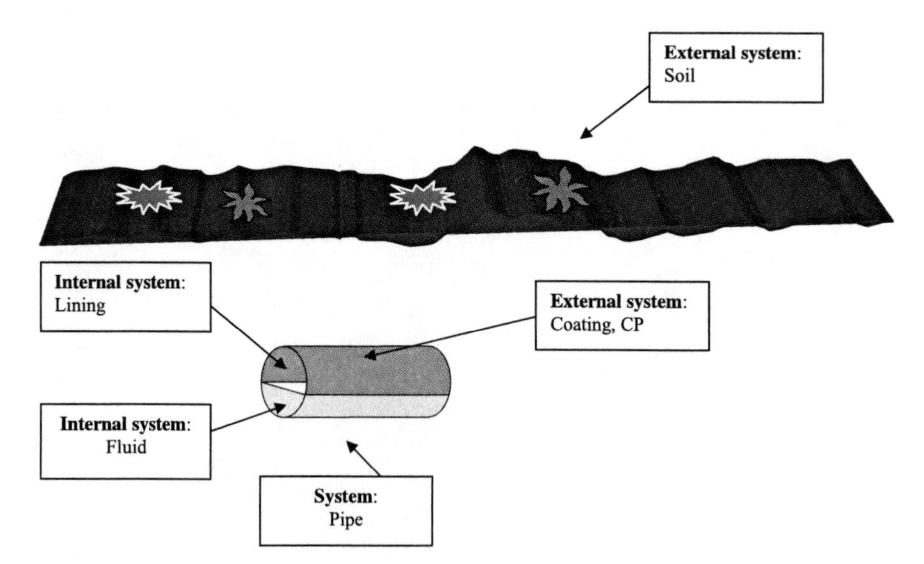

Figure 7.1 Corrosion systems and subsystems in a buried pipe [1]

As far as corrosion-related bacteria are concerned, in this chapter we will focus on both external and internal corrosion systems of a buried metallic pipeline.

A very important point, however, is the apparent discrepancy between our use of the term "external corrosion" and the way that it is addressed in corrosion literature, where most of the time, when external corrosion of buried pipes is mentioned, the damage to the coating and the "exterior" wall of the pipe is meant. As the reader can easily understand, our use of the term "external corrosion" includes this type of corrosion classification too. Therefore, we may use these terminologies interchangeably, bearing in mind that defining the surroundings of a buried pipe as the external system of corrosion will define a wider domain than just addressing what happens on the exterior wall of the pipe.

A study of failures of on-land oil and gas pipelines from 1970 to 1984 showed that more than 16% of the damage was due to corrosion, with 40% of it being external corrosion, and 17% internal corrosion [2]. Jack *et al.* also reported that the primary mechanism of deteriorating pipeline integrity was external corrosion of the buried pipes [3]. It is a common practice to address coating and CP as measures of protecting underground pipelines from "the effects of the environment" [4]. However, while external corrosion has been reported to be the main cause of underground pipe failures, the authors of a study regarding the share of the contribution of chemical, microbial and cathodic protection factors (such as the pipe-to-soil potential) to the underground corrosion of steel in anaerobic environments concluded that the microbial factor was the most important element [5].

Biofilms are reported to mainly form on the bottom of the internal surface of pipelines (over a sector of approximately 30° angle [6]), making them different from scale and corrosion products that are, for instance, generated over the whole surface in injection water pipelines [7].

Figure 7.2 summarises some of the most well-known failure mechanisms in buried pipelines. Some of these mechanisms have been explained and discussed in previous chapters in this book, such as the effect of hydrotesting on MIC (Chapter 5).

It must be noted, however, that while MIC could be an initiator of corrosion, it could well be a result as well. For instance, as shown in Figure 7.2, if the line is passing through different soils where the difference in the average diameter of the soil particles will allow different oxygen ingress gradients to be formed, this may increase the possibility of having differential aeration cells formed on the exterior wall of the underground pipeline. If, also, the coating is performing poorly, then, due to coating disbonding some areas with poor or no coating cover (holidays) are formed. Chances are that these holidays will be the best spots at which electrochemical corrosion starts. Being exposed to the community of the soil micro-organisms, including SRB and SOB, a "sulphureta"[1] may be created, depending on many factors (including weather conditions) as will be addressed later in this chapter.

[1] "Sulphureta" is a term used to address alternating oxidised and reduced sulphur environments, such as a bacterial consortia containing SRB (that reduce sulphur compounds) and SOB (that oxidise sulphur compounds). See [19].

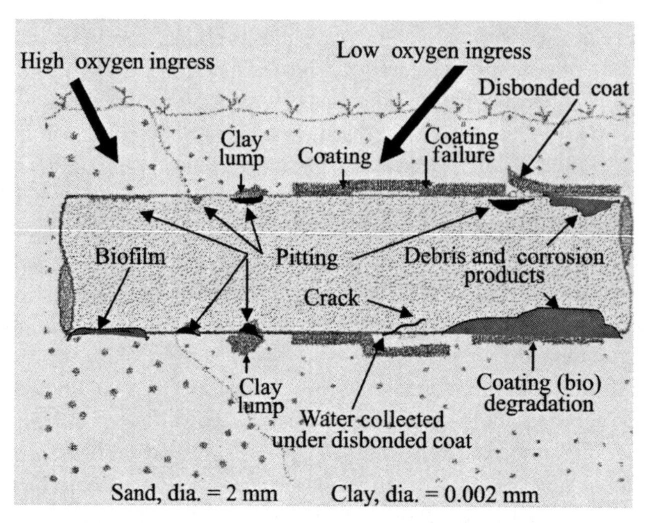

Figure 7.2 A review of some of the factors contributing to corrosion and particularly, MIC [8]

When condensed moisture and water are collected under disbonded coatings, at least two scenarios can occur:

1. The collected water is quite conductive.
2. The collected water is not very conductive.

If the trapped water has good conductivity and the pipeline is under CP, this will allow the current to pass and the required potential to be established so that the steel under the disbanded coat may be protected [9]. However, one should not forget that this water with relatively good conductivity is also a good electrolyte, thus raising the possibility of electrochemical corrosion under the disbonded area. On the other hand, if the trapped water is not a good conductor, the CP criteria will not have the opportunity to be maintained. If the water trapped under the coat is saturated with cations such as calcium or carbonate ions, making it quite alkaline, scaling may occur, and this plus an elevated pH may protect the underlying steel [3].

It may also be interesting to know that if the soil around the pipe contains SRB and SOB, as these two almost always accompany each other [10], they can work in "shifts" so that when the environmental conditions are suitable for the aerobic SOB – such as dry soil where inter-particle spaces and cavities are filled with oxygen – the SRB will wait, for example, until in the wet soil resulting from a rainy day, the oxygen trapped in the inter-particle spaces is expelled. This makes the environment so low in oxygen that the SRB can start to proliferate. This coexistence can enhance the corrosion even further.

7.3 Maritime Piled Structures (Jetty and Wharves)

A commonly seen problem with steel piles in ports and jetty structures is a type of electrochemical corrosion called "accelerated low water corrosion (ALWC)". An integral part of ALWC could be MIC [11].[2] In fact, some definitions of ALWC do consider MIC an integral part of the definition [12]. This type of corrosion has been observed and reported in ports all around the world, including the USA [13], Europe [12], and Australia [14]. In many cases of ALWC, microbial corrosion manifests itself as a mass which is orange in colour and collectively referred to as "orange bloom" (Figures 7.3a–d).

In essence, orange bloom can be regarded as a microbial community where SRB are definitely a part, due to the black iron sulphide mass associated with the orange bloom (Figure 7.4). Upon removal of the orange bloom, the liberated hydrogen sulphide produced by the SRB and the remaining black iron sulphide products can be detected. Orange bloom is capable of flagging very serious pitting of the steel piles and thus endangering their mechanical integrities (Figure 7.5).

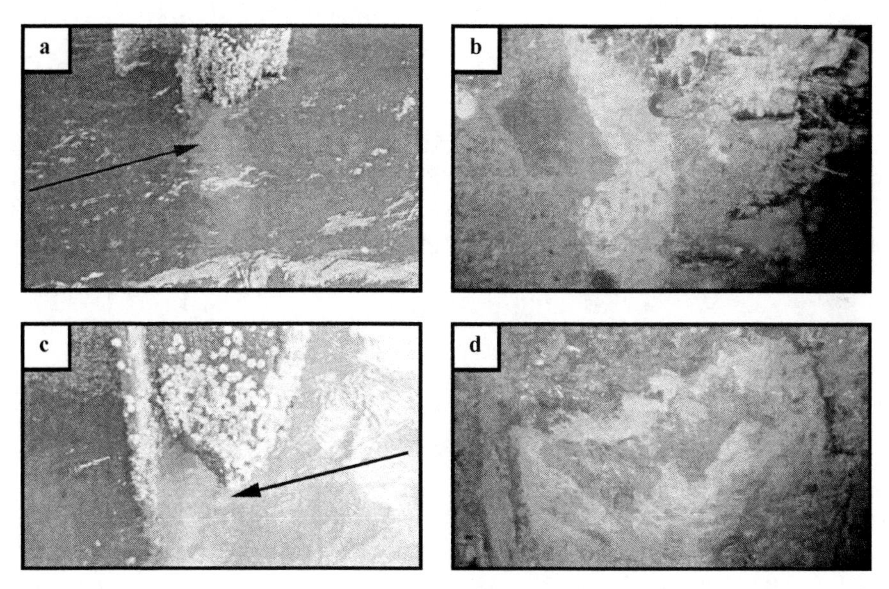

Figure 7.3 **a** "Orange bloom" (arrowed) as seen from above the water level. **b** Close-up of the same mass under water. **c** The steel underneath the orange bloom, and **d** its close-up after removal of the orange bloom (all images courtesy of Extrin Consultants)

[2] At a conference on Durability of Steel Pilings in Soil and Marine Environments in 1984, it was reported that "bacterial corrosion of steel piling in marine environments was not significant and … marine fouling appeared to be mostly beneficial". See [16] for more details.

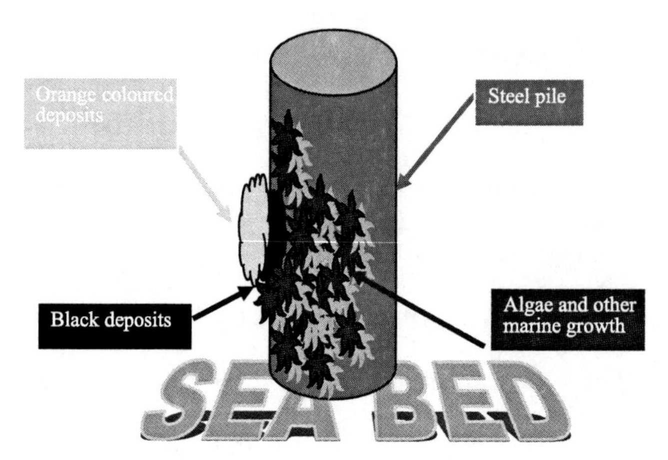

Figure 7.4 Schematic presentation of orange bloom on a steel pile [15] (not to scale)

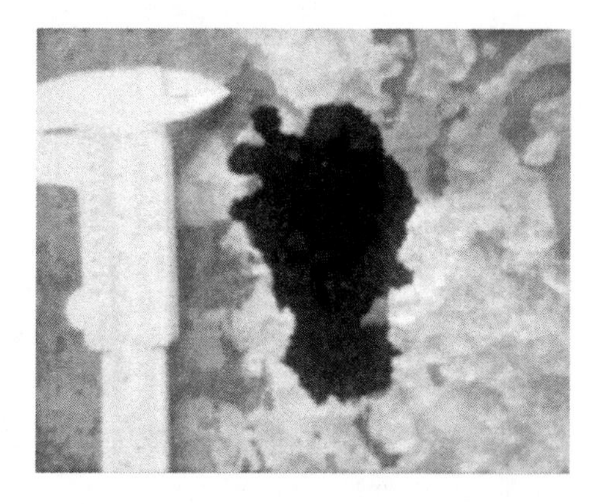

Figure 7.5 Perforation on the steel under the orange bloom (courtesy of Extrin Consultants)

There are still debates about the exact mechanisms that could be operative in ALWC. However, the involvement of bacterial species such as SRB and SOB has always been reported [16]. Figure 7.6 shows the factors that are important in ALWC and its occurrence.

An accepted scenario on the effect of SRB and SOB [17] is shown schematically in Figure 7.7. As is seen in the figure, when there is high tide and thus limited or no oxygen available, the anaerobic SRB will be able to use the anaerobic environment thus produced and reduce sulphates to sulphide that, when taken into consideration with the anodic reaction of dissolving iron and availability of iron ions, iron sulphides will be produced (thus the black colour of the "orange" bloom, see Figure 7.4). When, however, there is a low tide, oxygen becomes available to the sulphur-oxidising bacteria (SOB), where these bacteria are capable of using the situation to produce acidic conditions and very low pH.

By comparing Figure 7.7 with Figure 7.8, a general pattern may be reached: a cyclic corrosion effect of which SRB are an important part helps in intensifying corrosion in an environment that also contains SOB.

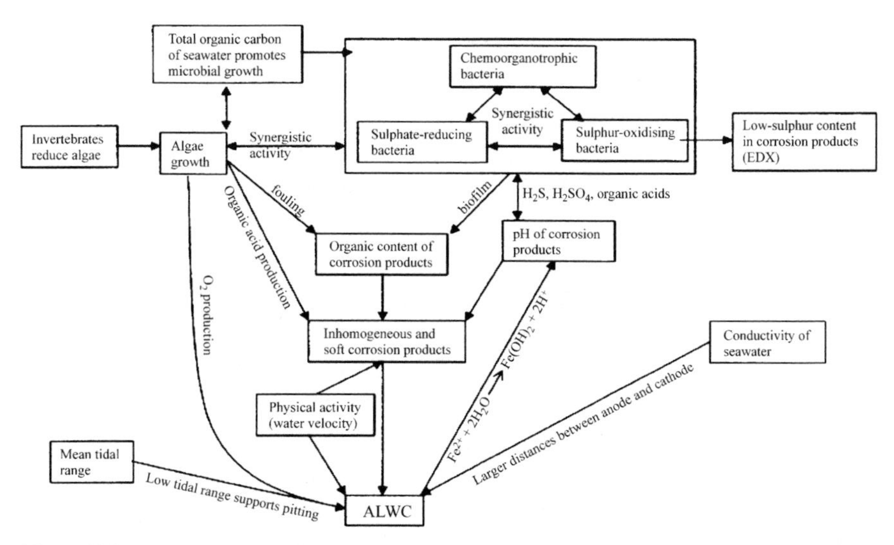

Figure 7.6 A presentation of factors that can be important in ALWC [16]

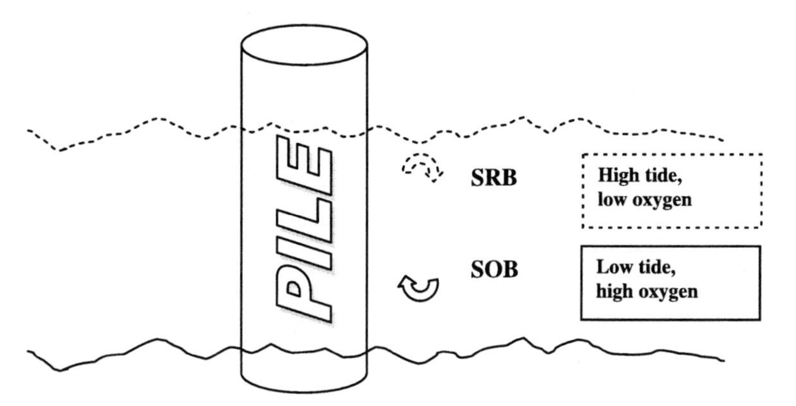

Figure 7.7 Possible cyclic effect of SRB and SOB on ALWC of steel piles

To prevent ALWC, the use of coatings such as coal tar epoxy or glass flake composite along with application of CP has been recommended [16]. Although replacing piles which are beyond economic repair can always be an option, the repair techniques that can normally be applied are one or a combination of the following [18]:

- Welding patch plates (for small areas showing signs of MIC)
- Welding strengthening plates (for areas where the effect of corrosion is more extensive, either U-plates or profiled plates can be used for the damaged areas)
- Plating with reinforced concrete infill (especially in Z-sections)
- Concrete as *in situ* collars or reinforced concrete plugging (*e.g.*, as encasement on H-piles)
- Splicing: it may be possible to cut out damaged sections of single piles (H, box or tubular) and join the replacement sections.

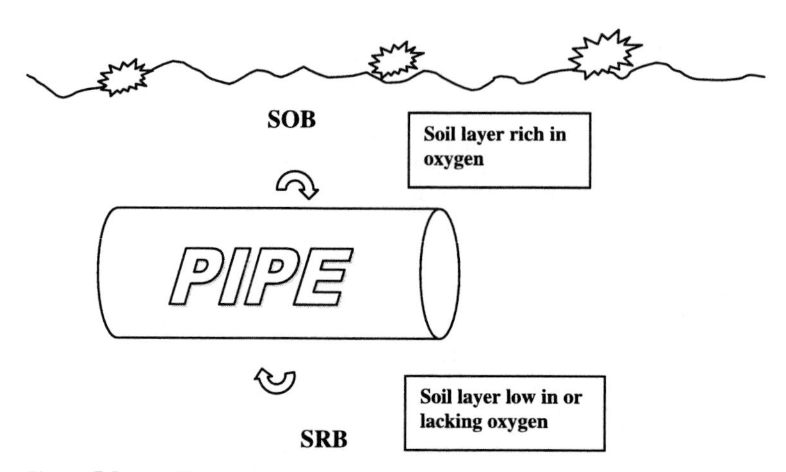

Figure 7.8 Possible cyclic effect of SRB and SOB on a buried pipe

7.4 Offshore Platforms

Offshore platforms are, in essence, similar to buried pipelines because in both, external and internal surfaces are exposed to corroding environments: in buried pipelines the external surface of the pipe is exposed to the soil (which is a corrosive environment), and its internal surface is under the corrosive impact of the fluid that is going through, either water, oil or the like. In case of offshore platforms, the whole immersed structure is exposed to seawater (a corrosive medium), and the internal surfaces of the systems such as seawater injection systems or oil storage facilities can be considered locations at which corrosion is occurring internally.

While there could be many ways to classify off-shore platforms and structures, one approach to address the basic types of offshore platforms (or, alternatively within the context of this book, offshore drilling units) is as follows:

1. Fixed platforms
2. Submersibles
3. Semi-submersibles
4. Jack-ups
5. Drilling ships
6. Tension-legs platforms

In an offshore platform, most of the MIC problems may happen in the following spots [19], some of which are shown in Figure 7.9:

- Marine fouling
- Drill cuttings around the platform legs
- Oil storage and transport
- Water-filled legs

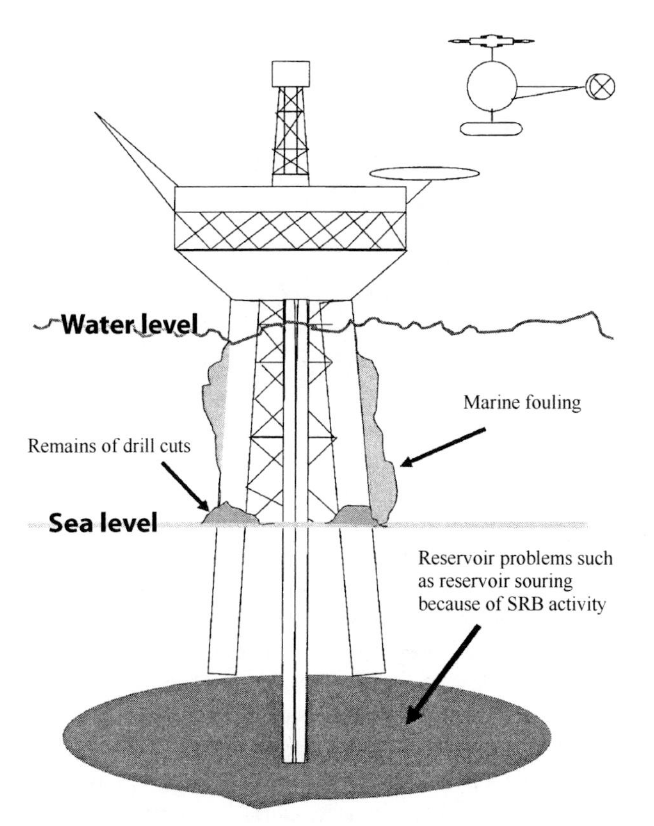

Figure 7.9 Some locations vulnerable to MIC in off-shore platforms[3]

- Production system
- Seawater injection system
- Downhole pipework
- Reservoir problems

In addition to a series of problems,[4] two main problems resulting from bacterial growth in offshore structures are [20]: (1) hydrogen sulphide production (generated, for example, by SRB) that besides being volatile and toxic, thus serious to the personnel safety, causes corrosion and souring of the products (crude oil, for instance) which ultimately affects the quality and final price [21], and (2) the production of bacterial metabolites which could give rise to accelerated materials deterioration.

---

[3] The figure has been taken from www.consrv.ca.gov/dog/picture_a_well/offshore_platform.htm with some modifications for our purpose here.

[4] Some of such problems are reservoir souring and/or filter blockages in diesel systems (communication with Dr. A.M. Orshed, Production Services Network, Aberdeen, UK, 01/June/07).

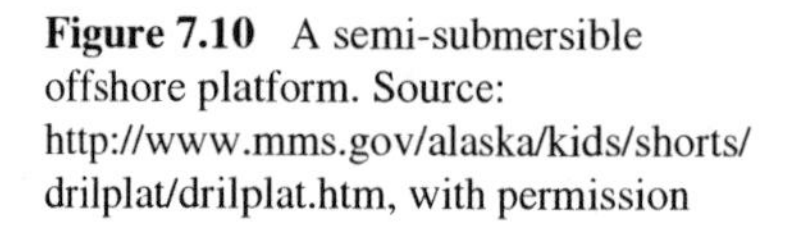

Figure 7.10 A semi-submersible offshore platform. Source: http://www.mms.gov/alaska/kids/shorts/drilplat/drilplat.htm, with permission

The effects on offshore structures can be categorised as external effects (such as environmental effects on external surfaces of these structures) and internal effects (such as MIC problems in water handling system and oil production system).

The impact of MIC on the internal systems is more apparent and immediate than those on the external surfaces. Careful monitoring, regular maintenance, and prudent use of biocides (such as chlorine and chlorine-releasing compounds, phenolics, aldehydes, and quaternary ammonium compounds) are some of the practices that can be recommended.

The type of MIC-related problems that may be expected in, for example, submersible and semi-submersible platforms are, more or less, similar to stagnant water problems caused in firewater lines or pipelines. Such platforms have pontoons and columns that, when flooded with seawater, cause the pontoons to submerge to a predetermined depth. Figure 7.10 shows an example of a semi-submersible offshore platform.

In such platforms, the stagnant water becomes de-aerated and oxygen-free, so it becomes a good place for SRB to become active. In addition, it was reported that the decay of macro-organisms in sea water, in the presence of light, could encourage phototrophic sulphur bacteria's role in increasing anaerobic corrosion of metals [22]. A routine countermeasure is the use of biocides. However, the biocide should not simply be introduced into the platform leg as it takes a substantial time (perhaps many weeks) to distribute. Real-life experiences [23] have shown that it will be necessary to overdose the biocide if a long lifetime is expected.

7.5 Firewater Lines

In an emergency when water is required to extinguish a fire, having a reliable firewater system that can handle stagnant water is of vital importance. Figures 7.11a and b show two examples of MIC within such systems.

It has been reported that the main cause of pitting in fire sprinkler systems is oxygen [25], as these systems use locally supplied potable water. This water,

a

b

Figure 7.11 **a** Internal wall of a firewater pipeline in which water has been stagnant and **b** observed perforation on the pipe (arrowed) [24]

being rich in oxygen, can easily establish differential aeration cells along the piping. It is then possible for oxygen to diffuse out of the system; rendering it more anaerobic [20]. Interestingly, an investigation has detected aerobic types of corrosion-enhancing bacteria, such as iron-oxidising bacteria (IOB), for example, *Gallionella* and *Siderophacus* which may also contribute to the corrosion seen in failed fire protection systems [26]. Therefore, it may be assumed that a cyclic action of anaerobic SRB and aerobic IOB such as anaerobic SRB and aerobic SOB in buried pipelines and steel piles (ALWC) could also be operational in firewater systems.

7.6 Summary and Conclusions

No matter how different two systems may seem at first sight, when it comes to MIC, some general patterns of corrosion can be recognised for both. Patterns such as those operative in a buried pipeline may well be similar to those of a submerged steel pipe and its ALWC problem. The stagnant water and the type of problems it produces have the same mechanisms for the involvement of MIC, whether in a firewater pipe or the water-filled legs of an offshore platform.

MIC may seem to be difficult to explain but, if recognised promptly and accurately, could have simple general patterns to look at, for both mitigation and prevention.

An integral part of the evaluation and assessment of the severity of MIC rests on the type of material that has been used. Examples of some materials that are frequently used in industry will be the topic of the next chapter.

References

1. Javaherdashti R, Marhamati EG (2005) A computerized model incorporating MIC factors to assess corrosion in pipelines. Mater Perform 44(1):56–59
2. Eiber RJ, Jones DJ, Kramer GS (1992) Analysis of DOT-OPSR data from 20-day incident reports 1970–1984. As quoted in Potts AE (1992) Accident analysis and reliability of offshore pipelines. Monash University Offshore Engineering Program
3. Jack TR, Van Boven G, Wilmott M, Sutherby RL, Worthingham RG (1994) Cathodic protection potential penetration under disbonded pipeline coating. Mater Perform 33(8):17–21
4. Touzet M, Lopez N, Puiggali M (1999) Effect of applied potential on cracking of low-alloyed pipeline steel. in low pH soil environment. In: Advances in corrosion control and materials in oil and gas production (EFC 26). Jackman PS, Smith LM (eds) Woodhead Publishing
5. Li SY, Kim YG, Kho YT (2003) Corrosion behavior of carbon steel influenced by sulfate-reducing bacteria in soil environments. Paper No. 03549. CORROSION 2003, NACE International, USA
6. King RA (2007) Trends and developments in microbiologically induced corrosion in the oil and gas industry. MIC – An International Perspective Symposium. Extrin Corrosion Consultants, Curtin University, Perth, Australia, 14–15 February 2007
7. King RA (2007) Microbiologically induced corrosion and biofilm interactions. MIC – An International Perspective Symposium. Extrin Corrosion Consultants, Curtin University, Perth, Australia, 14–15 February 2007
8. Javaherdashti R (2000) A review of microbiologically influenced corrosion of buried cathodically protected coated gas pipe lines. In: Persian Department of Technical Education, Iranian National Gas Company, Tehran, Iran
9. Jack TR, Wilmott MJ, Sutherby RL (1995) Indicator minerals formed during external corrosion of line pipe. Mater Perform 35(11):19–22
10. Tatnall RE (1993) Introduction. In: A practical manual on microbiologically influenced corrosion Kobrin G (ed), NACE, Houston, TEXAS USA
11. Javaherdashti, R. (2006) Microbiological Contribution to Accelerated Low Water Corrosion of Support Piles, *Port Technology International,* pp. 59–61, 29th Edition
12. Gehrke T, Sand W (2003) Interactions between micro-organisms and physicochemical factors cause MIC of steel pilings in harbours. (ALWC) Paper No. 03557.CORROSION 2003, NACE International, Houston, Texas USA
13. Hannam MJ, Clubb DL (2002) Experience and considerations on the corrosion protection of harbour steel sheet piling. The Institute of Corrosion Conference, Cardiff, UK, 23 October 2002
14. Hutchinson C, Farinha PA, Vallini D (2004) The effectiveness of petrolatum tapes and wraps on corrosion rates in a marine service environment. Paper No. 033. Corrosion and Prevention 2004 (CAP04) Perth, Australia, 21–24 November 2004
15. Javaherdashti R (2005) Microbiological contributions to accelerated low water corrosion (ALWC) of steel-piled structure: A review. Proceedings of Corrosion and Prevention 2005 (CAP05) Gold Coast, Australia, November 2005
16. Gubner R, Beech I (1999) Statistical assessment of the risk of the accelerated low-water corrosion in the marine environment. Paper No. 318. CORROSION–99, NACE International, USA
17. Little B, Lee JJ, Ray R (2007) How marine conditions affect severity of MIC of steels. MIC – An International Perspective Symposium. Extrin Corrosion Consultants, Curtin University, Perth, Australia, 14–15 February 2007
18. Christie J (2007) Dealing with MIC on maritime piled structures. MIC – An International Perspective Symposium. Extrin Corrosion Consultants, Curtin University, Perth, Australia, 14–15 February 2007
19. Edyvean RG, Dexter SC (1993) MIC in marine industries. In: A practical manual on microbiologically influenced corrosion Kobrin G (ed) NACE, Houston, Texas USA

20. Wilkinson TG (1983) Offshore monitoring in microbial corrosion: Proceedings of the conference sponsored and organised jointly by The National Physical Laboratory and The Metals Society, 8–10 March 1983, The Metals Society, London
21. Evans P, Dunsmore B (2006) Reservoir simulation of sulfate-reducing bacteria activity in the deep sub-surface. Paper No. 06664. CORROSION-2006, NACE International, USA
22. Eashwar M, Maruthamthu S, Venkatakrishna Iyer S (2004) A possible role for phototrophic sulphur bacteria in the promotion of anaerobic metal corrosion Current Science 86(5):639-641
23. NACE CORROSION NETWORK (2002) Discussion Group: Corrosion within offshore jacket legs
24. Fernance N, Farinha PA, Javaherdashti R (2007) SRB-assisted MIC of fire sprinkler piping. Mater Perform 46(2)
25. Brugman HH (2004) Corrosion and microbiological control in fire water sprinkler systems. Paper No. 04512. CORROSION-2004, NACE International, USA
26. Yee GG, Whitbeck MR (2004) A microbiologically influenced corrosion study in fire protection systems. Paper No. 04602. CORROSION-2004, NACE International, USA

Chapter 8
Examples of Some Materials Vulnerable to MIC

8.1 Introduction

Without a doubt, the choice of material is an important factor to make a system resistant or vulnerable to MIC. Case histories show that carbon steel is a more susceptible material in comparison with stainless steels, and that stainless steel SS316 is more resistant than SS304.

This chapter will focus on three types of materials: duplex stainless steel, copper and copper-nickel alloys, and concrete. The main reason for selecting these materials was that they are of frequent use in industry. For example, copper and copper alloys have a reputation that no micro-organism can colonise them, as copper is poisonous to living organisms. This "copper reputation" has given this material a very wide range of applications. Duplex stainless steels are better known for their upgraded corrosion resistance versus the "ordinary" stainless steels such as grades 316 and 304 and their varieties.

On the other hand, concrete, thanks to its composite structure that takes advantage of both steel and cement, has an impossible-to-ignore position among other materials, especially in the sewage treatment industry.

We start this chapter with copper alloys, as none of other materials have the so-called "bio-resistance" of copper.

8.1.1 Copper and Cupronickels

Localised corrosion of copper can occur in four types, as summarised and addressed by Yakubi and Murakami [1] and tabulated in Table 8.1.

It was "known" for quite some time that the copper sheets that had been used to cover the bottoms of wooden ships corroded in seawater such that the environment could be kept toxic to barnacles and similar organisms [2], and thus biofouling-

Table 8.1 Classification of copper corrosion types in water

Type	Water type	pH range	Water temperature	Features
I	Hard	7–7.8	Cold	Not reported
II	Soft	Below 7.2	Hot	Deep, narrow pit morphology and existence of a basic copper sulphate product
III	Soft	Above 8.0	Cold	Wide and shallow pit morphology, evidenced by production of "blue water", and pipe blockage
Moundless	Containing high sulphate ion and silicon dioxide	Not reported	Not reported	Open-mouth pit morphology, no "mounds" of corrosion products present on such pits

free. Copper and copper alloys are still praised[1] today for their resistance to biocorrosion.

However, the involvement of some types of micro-organisms with relatively high tolerance to copper has been reported. In their review of the behaviour of cupro-nickels alloys in sea water, Parvizi *et al.* [3] reported *Thiobacillus thiooxidans* being able to withstand copper ion (cuprous) concentrations as high as 20,000 ppm. Palanichamy *et al.* [4] also observed endospore-forming genus *Bacillus* and non-endospore forming genus *Propionibacterium* on copper surfaces. Critchley *et al.* [5] have reported the isolation of copper-resistant species such as *Sphingomonas* and *Acidovorax*.

Microbial corrosion has been proposed as a possible cause for "blue water" corrosion [5]. Blue water corrosion is a term to address the release of copper corrosion by-products into the water, especially drinking water. It was reported that this type of copper corrosion has been most often observed when the water has been stagnant for several hours or days, and typically containing 2–20 ppm copper concentration (the recommended copper concentration in drinking waters is 2 ppm) [5]. Blue water corrosion generally occurs randomly. However, blue water has been reported not to significantly compromise the pipe integrity in general [6].

Two models can be proposed to explain MIC of copper. As Webster *et al.* [6] put it, these models can be explained as follows:

---

[1] See, for example, reviews by Schleich W, Steinkamp K, "Biofouling Resistance of Cupronickel-Basics and Experience", Paper No. P0379, Stainless Steel World, Maastricht, The Netherlands, 2003 and also Schleich W, "Typical Failures of CuNi 90/10 Seawater Tubing Systems and How to Avoid Them", Paper No. 12-0-124, EuroCorr 2004, Nice 2004. Also, Powell C, and Michels H, "Review of Splash Zone Corrosion and Biofouling of C70600 Sheathed Steel During 20 Years Exposure", EuroCorr 2006, Event No. 280, 24–28 September 2006, Maastricht, The Netherlands.

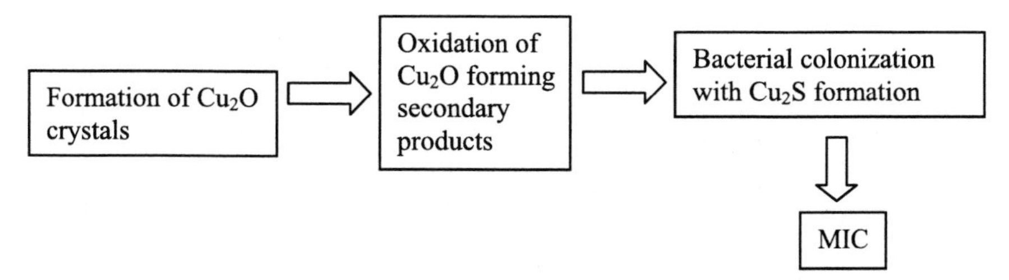

Figure 8.1 Possible MIC pattern for MIC of Cu-10%Ni in non-chlorinated brackish water according to de Romero [9]

Model I: The EPS (extracellular polymeric substances) – which is mainly the biofilm – create preferential cathodic sites by the "cation-selective nature of the EPS".[2]

Model II: This model describes MIC of copper in terms of the formation of a copper ion concentration cell by the EPS and the generation of a weakly acidic environment.

Webster *et al.* consider that the second model, which is based on a decrease in pH, is probably the prevailing mechanism.

Cupronickels (90/10 that contains 10% nickel or 70/30 with 30% nickel or Monel 400) have been used for many years in applications where sea water has been involved, for their good corrosion resistance. This fitness for purpose is specifically because of the cupronickels' passive cuprous oxide (Cu_2O) film, which retards both the anodic dissolution of the alloy and the rate of oxygen reduction [7]. Based on studies by Gouda *et al.* and reported by Lee *et al.* [8], alloy 400 (= Monel 400 containing 66.5% nickel, 31.5% copper, and 1.25% iron) is much more susceptible to SRB-induced MIC compared to 70/30 cupronickel or brass.

De Romero *et al.* [9] also suggested patterns as possible mechanisms for MIC of Cu-10% Ni in non-chlorinated brackish water, where because of a lack of chlorine and the possibility of surviving micro-organisms, MIC is possible. Their proposed mechanism for MIC of Cu-10% Ni in brackish water with no chlorine is schematically summarised in Figure 8.1.

8.1.2 Duplex Stainless Steels

Carbon steel and stainless steels and their behaviour with regard to microbial corrosion have been relatively well studied and documented compared to duplex stainless steels.

Duplex stainless steels (DSSs) are being used in many industries, such as chemical processing, electrical energy generation [10] and the oil and gas industries,

---

[2] Biofilms are negatively charged.

where they are susceptible to corrosion (mainly SCC) in environments such as Packer fluid and acidising fluid [11].

DSSs have two phases, austenite and ferrite, where their presence and particular ratio influence the way these steels interact with the environment [12]. An example of a typical microstructure of a duplex stainless steel is shown in Chapter 9. The austenite phase provides features such as toughness and weldability, and the ferritic phase contributes to strength, corrosion resistance, and SCC resistance [13].

The probability of chloride SCC in some DSSs known as SAF2205 has been reported at less than 10% [14]. However, when hydrogen sulphide is present in the environment, the danger of hydrogen-assisted chloride SCC for DSSs increases with temperatures in the range of 60°C–100°C and decreases with higher Cr, Mo, and N contents [11].

The mechanisms regarding DSS characteristics of corrosion resistance are still not well understood. Some of the theories in this regard are (a) a combined effect of corrosion potentials in each phase and its impact on crack initiation and propagation in either austenite or ferrite or both phases [15], (b) a difference in the potential of grain boundaries relative to the ferrite *per se* [16], and (c) a mechanical effect of austenite and ferrite and the impact of hydrogen diffusion in ferrite to compensate for the produced stresses [12].

Duplex stainless steels are also vulnerable to microbial corrosion; SAF 2205 has been reported as being vulnerable to MIC [17, 18], particularly in the presence of SRB [19, 20].

The vulnerability of DSSs to MIC is important, as it once again proves that just by increasing some alloying elements that have a reputation for inducing corrosion resistance, such as chromium, one cannot overcome MIC. To avoid the risk of MIC, a careful material selection must be accompanied by scrutiny of the service conditions and a serious follow-up that monitors how the material is performing.

8.1.3 Concrete

As Rogers *et al.* [21] quoted to The U.S. Nuclear Regulatory Commission, "Service Life of Concrete", complied in 1989, there are at least seven major chemical/physical factors reported to be major causes of concrete degradation:

1. Sulphate and chloride attack
2. Alkali aggregate reactions
3. Water leaching
4. Freeze/thaw cycling
5. Salt crystallisation
6. Corrosion with resulting expansion of reinforcing bars
7. Acid rain

As seen, biodegradation of concrete is not among these causes. This is an example of how authorities can be oblivious to the biodeterioration of concrete. As

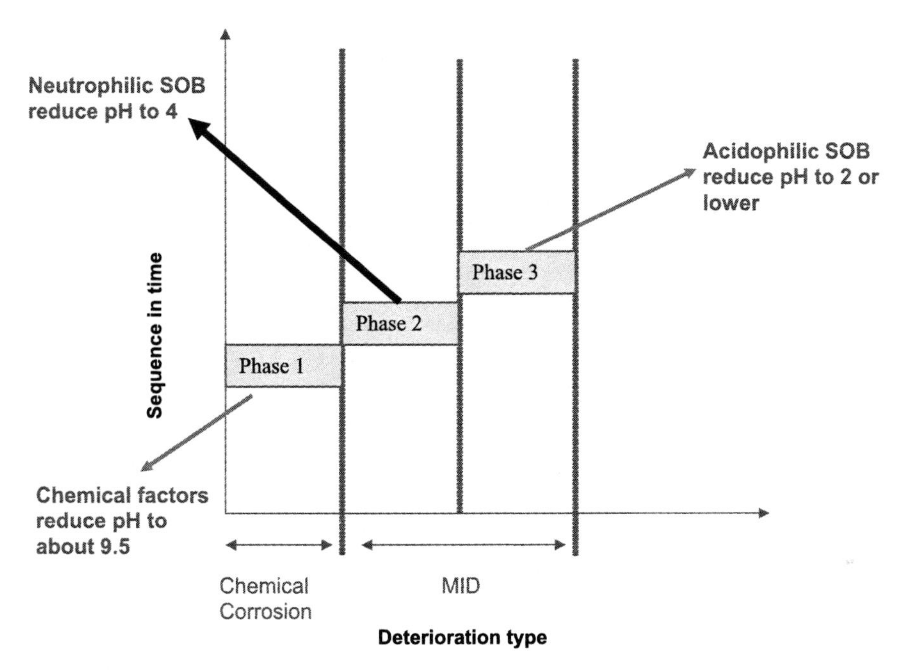

Figure 8.2 Schematic summary of deterioration of concrete with the involvement of MID [26]

Rogers *et al.* put it: "an understanding of concrete degradation may be incomplete without including the effects of microbial influenced degradation, or briefly, MID".[3]

There are case histories [22, 23] reporting SRB-induced infection of the concrete columns (up to 70% in some areas) of an occupied building. What is thought to be the main mechanism for attacking concrete itself is by the act of SOB bacteria such as *Thiobascillus thiooxidans* that excrete very low pH acid (H_2SO_4) which dissolves the concrete [24]. In sewer pipes, SOB can contribute to corrosion rates of up to 1 cm/year [25].

More precisely, it is a process that can be schematically shown (Figure 8.2). The MID-assisted deterioration of concrete can happen in three phases. To date, nothing is known regarding the time intervals between each step, but it seems that the concrete becomes vulnerable first by chemical corrosion (deterioration) because of factors such as the formation of carbonic acids. This will lower the pH from above 12.0 to somewhere around 9.0–9.5. Then "microbial succession" starts, where neutrophilic SOB are replaced by another group of SOB which are capable of further reducing the pH, thus dissolving the concrete.

MID can be seen as a three-phase process whose phases are schematically summarised in Figure 8.2.

---

[3] Corrosion, and thus MIC, is used to address degradation in metals. We will use the term "microbial influenced degradation", or briefly MID, to address degradation of non-metallics.

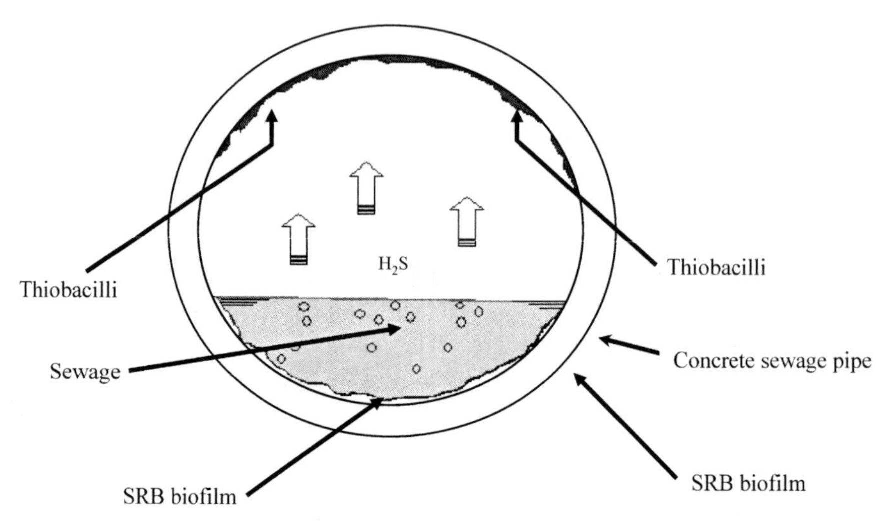

Figure 8.3 Schematic representation of possible microbial consortium in a concrete sewage pipe [29]

As seen in Figure 8.2, the three phases can be explained as follows [27]:

- Phase 1: Combined corrosive effects of atmospheric carbon dioxide and hydrogen sulphide reduce the pH to about 9.5.
- Phase 2: First stage of "microbial succession" where, provided that sufficient nutrients, moisture and oxygen exist, some species of sulphur-oxidising bacteria (*e. g.*, *Thiobacillus sp.*) can attach themselves onto the concrete surface and grow. Mostly, these species of SOB are neutrophilic sulphur-oxidising bacteria (NSOM). These bacteria produce some acidic products and convert the sulphides present to elemental sulphur and polythionic acids.
- Phase 3: The second step of microbial succession, it normally follows Phase 2 where the pH has been reduced fairly. Another species of SOB known as acidophilic sulphur-oxidising bacteria (ASOM) such as *T. thiooxidans* colonise the concrete surface and further reduce the acidity. It has been proposed that during Phase 2 the NSOM reduces the pH to 4.0 where during Phase 3, the pH is further reduced by the ASOM to 1.0 or 2.0 [28].

Studies show that microbial succession can start with very low numbers of both types of the sulphur-oxidising bacteria so that MID can develop completely [27]. Quoting from Bock and Sands' work, Rogers *et al.* reported that a cell density of chemolithotrophic SOBs such as *Thiobacillus* of about 10^4 to 10^6 cells per grams of concrete is required before MID is detected [27].

When concrete is used in environments such as sewer systems, it can be exposed to a cyclic action of SRB and SOB (Figure 8.3), in a sense, similar to ALWC (Chapter 7). In this way, SRB and SOB will have a synergistic effect on each other in terms of enhancing corrosion (Figure 8.4).

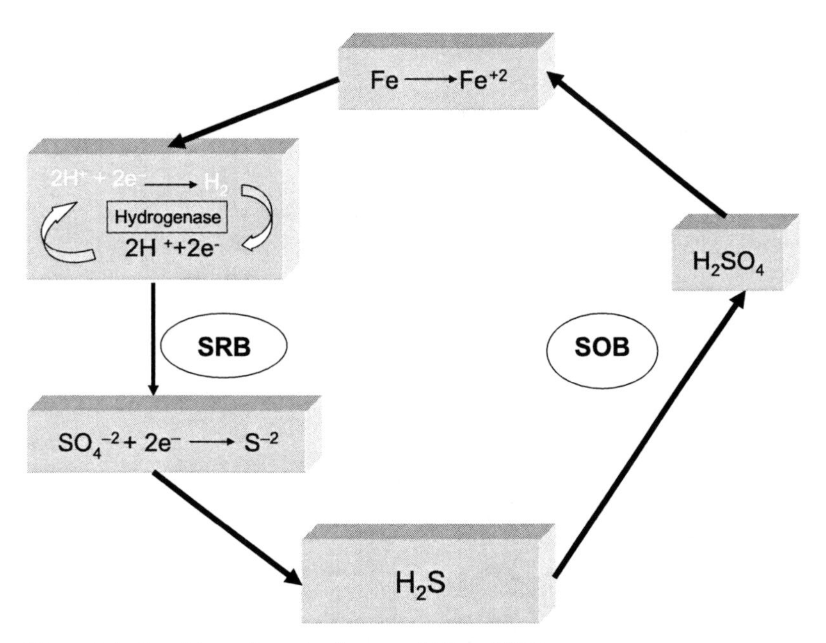

Figure 8.4 Possible cyclic action between SRB and SOB

At low sulphate ion concentrations (less than 1000 ppm), the corrosion product is ettringite (3CaO. Al_2O_3. $CaSO_4$. 12H_2O or 3CaO. Al_2O_3. 3$CaSO_4$. 31H_2O), whereas gypsum ($CaSO_4$. 2H_2O) is the main cause of deterioration at high sulphate ion concentrations. It follows then, that the mechanism of attack depends on the concentrations of the SO_4^{2-} ions in the solution [30]. It has been reported that ettringite is produced when the pH levels are higher than 3.0, whereas gypsum is likely to be formed at pH levels less than 3.0 [31]. Also, it must be noted that ettringite is expansive and causes internal cracking, which is actually providing a larger surface for chemical reactions to occur, thus resulting in more sites of penetration into the concrete [27].

The conversion of the concrete into gypsum and ettringite reduces the mechanical strength of the concrete, which is followed by a reduction of the structural integrity of the concrete and may result in total failure and collapse of the structure.

In their review [32], Ribas Silva and Pinheiro quote from the work done by Salvadori with regard to the impact of some biocides on several inorganic materials including concrete. This impact is tabulated in Table 8.2. In addition to chemical treatment of concrete with biocides, other techniques of dealing with concrete such as mechanical and biological measurements were reviewed [32].

Table 8.2 Impact of Some Common Biocides on the Bacteria within the Concrete

Biocide type	Action
Acids	Cell death
Alkalis	Breaking up the bacteria acting on organic matters
Oxidant agents	Release oxygen or other active compounds
Surface-active agents (detergents)	Cause loss of structural organisation of cellular membranes
Phenols	Effective on cellular membranes and bacterial walls
Heavy metals	Toxic actions on proteins
Alcohols	Cause dehydration
Nitrogen-organics	Interfere with photosynthesis
Phospho-organics	Interfere with the biosysnthesis of some aminoacids

8.2 Summary and Conclusions

In this section, copper and cupronickels, duplex stainless steels, and concrete and their vulnerability to MIC were briefly reviewed. We showed that all of these materials are actually susceptible to microbial corrosion, and that thus there is no material that can be regarded as being completely safe from MIC. Chapter 9 concentrates on the treatment of MIC.

References

1. Yakubi A, Murakami M (2007) Critical ion concentration for pitting and general corrosion CORROSION 63(3):249–257
2. Burns RM, Bradley WW (1967) Protective coatings for metals. American Chemical Society, Monograph Series, 3rd edn
3. Parvizi MS, Aladjem A, Castle JE (1988) Behaviour of 90–10 cupronickel in sea water. Intl Materials Rev 33(4):169–200
4. Palanichamy S, Maruthamuthu S, Manickam ST, Rajendran A (2002) Microfouling of manganese-oxidizing bacteria in tuticorin harbour waters. Current Science 82(7):865–86
5. Critchley M, Taylor R, O'Halloran R (2005) Microbial contribution to blue water corrosion Mater Perform 44(6):56–59
6. Webster BJ, Werner SE, Wells DB, Bremer PJ (2000) Microbiologically influenced corrosion in potable water systems – pH effects. CORROSION 56(9):942–950
7. Shalaby HM, Hasan AA, Al-Sabti F (1999) Effects of inorganic sulphide and ammonia on microbial corrosion behaviour of 70Cu–30Ni alloy in sea water. Brit Corrosion J 34(4):292–298
8. Lee JS, Ray RI, Little BJ (2003) A comparison of biotic and inorganic sulphide films on Alloy 400. Proceedings of Corrosion Science in the 21st Century, vol. 6, Paper C057, UMIST UK
9. de Romero M, Duque Z, de Rincon O, Perez O, Aranjo I, Martinez A (2000) Online monitoring systems of microbiologically influenced corrosion on Cu–10%Ni alloy in chlorinated brackish water. CORROSION 56(8):867–876

10. Chaves R, Costa I, de Melo HG, Wolynec S (2006) Evaluation of selective corrosion in UNS S31803 duplex stainless steel with electrochemical impedance spectroscopy. Electrochimica Acta (51):1842–1846
11. Rhodes RR, Skogsberg LA, Tuttle RN (2007) Pushing the limits of metals in corrosive oil and gas well environments. CORROSION 63(1):63–100
12. Archer ED, Brook R, Edyvean RGJ, Videla H (2001) Selection of steels for use in SRB environments. Paper No. 01261. CORROSION–2001, NACE International USA
13. Siow KS, Song TY, Qiu JH (2001) Pitting Corrosion of Duplex Stainless Steels Anti-Corrosion Methods and Materials 48(1):31–36
14. Stainless Steel Selection Guide (2002) Central States Industrial Equipment & Service Inc. http://www.al6xn.com/litreq.htm USA
15. Gunn RN (1997) Duplex stainless steels. Ch. 7, Woodhead Publishing Ltd.
16. Danko JC, Lundin CD (1995) The effect of microstructure on microbially influenced corrosion Proceedings of International Conference on Microbiologically Influenced Corrosion, New Orleans, Louisiana, NACE International, USA, May 8–10, 1995
17. Kovach CW, Redmond JD (1997) High performance stainless steels and microbiologically influenced corrosion wwwavestasheffieldcom acom 1–1997
18. Neville A, Hodgkiess T (1998) Comparative study of stainless steel and related alloy corrosion in natural sea water. Brit Corrosion J 33(2):111–119
19. Johnsen R, Bardal E (1985) Cathodic properties of different stainless steels in natural seawater. CORROSION 41(5):296–302
20. Antony PJ, Chongdar S, Kumar P, Raman R (2007) Corrosion of 2205 duplex stainless steel in chloride medium containing sulphate-reducing bacteria. Electrochimica Acta (52):3985–3994
21. Rogers RD, Knight JJ, Cheeseman CR, Wolfram JH, Idachaba M, Nyavor K, Egiebor NO (2003) Development of test methods for assessing microbial influenced degradation of cement-solidified radioactive and industrial waste. Cement and Concrete Res (33):2069–2076
22. Scott PJB, Davies M (1992) Microbiologically-influenced corrosion Civ Eng 62):58–59
23. Davies M, Scott PJB (1996) Remedial treatment of an occupied building affected by microbiologically influenced corrosion Mater Perform 35(6):54–57
24. Little BJ, Ray RI, Pope RK (2000) Relationship between corrosion and the biological sulfur cycle: A review. CORROSION 56(4):433–443
25. Knight J, Cheeseman C, Rogers R (2002) Microbial influenced degradation of solidified waste binder. Waste Management (22):187–193
26. Javaherdashti R, Farinha PA, Sarker PK, Nikraz H (2006) On microbial corrosion of concrete: Causes mechanisms and mitigation. Concrete in Australia 32(1)
27. Roberts DJ, Nica D, Zuo G, Davis JL (2002) Quantifying microbially induced deterioration of concrete: Initial studies. International Biodeterioration & Biodegradtion (49):227–234
28. Davies JL, Nica D, Shields K, Roberts DJ (1998) Analysis of concrete from corroded sewer pipe. International Biodegradation & Biodegradation (42):75–84
29. Javaherdashti R (2004) A review of microbiologically influenced corrosion with emphasis on concrete structures. Proceedings of Corrosion and Prevention 2004 (CAP04), 21–24 November 2004, Perth, Australia
30. Monteny J, Vincke E, Beeldens A, De Belie N, Taerwe L, Van Germet D, Verstraete W (2000) Chemical microbiological and in situ test methods for biogenic sulfuric acid corrosion of concrete Cement and Concrete Res (30):623–634
31. Mori T, Nonaka T, Tazaki K, Koga M, Hikosaka Y, Noda S (1992) Interactions of nutrients moisture and pH on microbial corroson of concrete sewer pips. Water Resources 26(1): 29–37
32. Ribas Silva M, Pinheiro SMM (2007) Mitigation of concrete structures submitted to biodeterioration. MIC – An International Perspective Symposium. Extrin Corrosion Consultants, Curtin University, Perth, Australia 14–15, 2007

Chapter 9
How Is MIC Treated?[1]

9.1 Introduction

No matter how good and reliable the techniques and methods are for defining, recognising and detecting MIC, all will become pointless if the problem cannot be cured.

Treatment programs can be divided in two: to mitigate an existing problem or to prevent the initiation of a problem, right from the beginning. For reasons which are beyond this book, and have been explained to some extent elsewhere [1], most of the time what is required is mitigation.

There are very innovative ways to deal with a biocorrosion problem. Davies and Scott explain a very interesting case involving the paint on many of the sheathings of the structural columns of a fully occupied university medical building [2, 3]. Shortly after the building opened, the paint started to blister and "bled a colourless liquid that quickly became rust-coloured". In fact, 45% of all the columns tested showed sign of corrosion. Despite many practical restrictions and limits, the investigators could isolate microbial consortia containing sulphate-reducing bacteria. The source of the problem was attributed to contaminated, untreated water that had been used in making the concrete. However, no conventional way of using biocides could be applied, due to the fact that the building was already occupied. As a biocide, the chemical that was selected and applied was denatured ethanol-based chlorhexidine digluconate, used in some mouthwashes to treat gum diseases.

Another example of such non-conventional, innovative methods is applying immunoglobulin solutions films on the surface of carbon steel and stainless steel that has been shown to prevent the adherence of *Pseudomonas fluorescens* on these metallic surfaces, thus inhibiting biofilm formation [4]. However, this method would not become popular in industry, perhaps because of reasons such as

---

[1] The title of this chapter should have been "How is MIC technically treated?" to also address the nontechnical CKM-related treatment of MIC. However, the application of CKM to MIC problems is not different from applying it to corrosion, either microbial or "non-microbial".

the relatively high cost of immunoglobulin and lack of communication between the involved disciplines.[2]

In addition, in nature there are mechanisms that many industrial biocidal treatments have imitated. W.F. McCoy has given some examples of such systems [5]. For example; when water reacts with chlorine, bromine, or iodine, hypohalous acids are formed. These acids are of biocidal use in industry. Equivalently, in nature, in addition to the human immune system, this acid is also produced on the surface of some aquatic plants, keeping them free from germs.

All the above examples can serve to show that the treatment of microbiologically influenced corrosion cases are not always expensive or environmentally unfriendly practices. With lateral thinking and multidimensional planning based on an understanding of the mechanisms of microbial corrosion, it is possible to make a change, when necessary.

Microbial corrosion can be treated four ways:

- Physical-mechanical
- Chemical
- Electrochemical
- Biological

This chapter explains some physical-mechanical treatments (such as UV and pigging), chemical treatments (use of biocides, the advantages and disadvantages of some biocides, and treatment regimes such as dual biocide treatment,) electrochemical treatments (use of cathodic protection and coatings), and some biological treatments that are currently being researched and applied.

9.2 Physical-mechanical Treatments

9.2.1 Pigging

A pipeline inspection gauge, or "pig", is a tool with which, among many other tasks and benefits, pipelines are cleaned and/or inspected internally. Figure 9.1a shows an example of a pig to be used for cleaning natural gas pipelines. Figure 9.1b illustrates the relative size of a pig.

Some of the reasons for running pigs are [8]:

- Improving the flow efficiency of the pipeline
- Improving or insuring that useful and good data is gained on inspection by running a pig
- Feedback on the results of chemical treatment programs that aid in increasing the service life of the pipeline
- Removing more debris and solid products

[2] Private communication with Professor Hector A. Videla, 15 August 2006.

a b

Figure 9.1 **a** An example of a "pig" used to clean natural gas pipelines [6]. **b** A pig being installed ino a part of a pipeline [7]

In addition, by running pigs, targets such as the removal of collected water and corrosion tubercles can be achieved if the facility (pipeline, for example) has been designed to allow pigging and is well-equipped with appropriate launchers and receivers [9]. The Verleun reference [8] reports some of the cases in which the problem of "unpiggable pipelines" has been dealt with successfully although factors such as the existence of over- or undersized valves, differently sized repair sections, and short-radius or mitred bends has caused no accessibility for pigging. The options to overcome the problem of dealing with a pipeline being unpiggable are (1) modification of the pipeline so that it becomes piggable (which is costly and may cause operation interruptions, and which in some cases is just impossible due to the nature of the performance), and (2) modification of the inspection equipment in accordance with the existing conditions [7]. This option means that, while there may be overlaps in terms of the design and operation of modified pigs, each case needs to be dealt with individually.

A closer look at Figure 9.1a may reveal the brushes around the main structure of the pig. Due to the extreme conditions of temperature, moisture, and mechanical abrasion to which pigs are exposed, these wire brushes start to degrade and also corrode. In the case of the so-called "intelligent pigs" that for their operation use "magnetic flux leakage" techniques, the necessary electrical contact between the pig and the wall of the pipe is provided by these steel brushes. Neither austenitic steels (as they lack ferrite, and thus cannot be magnetised) nor ferritic seels (due to their relatively low work hardening rates) can be used for making these brushes, leaving the door open to other types of steel such as martensitic steels and duplex stainless steels [10] (where both austenite and ferrite are present, Figure 9.2). Materials selection plays an important role here in terms of suggesting a material that can improve the performance of an inspection/cleaning tool and thus, perhaps indirectly, add to the increased life of the pipeline.

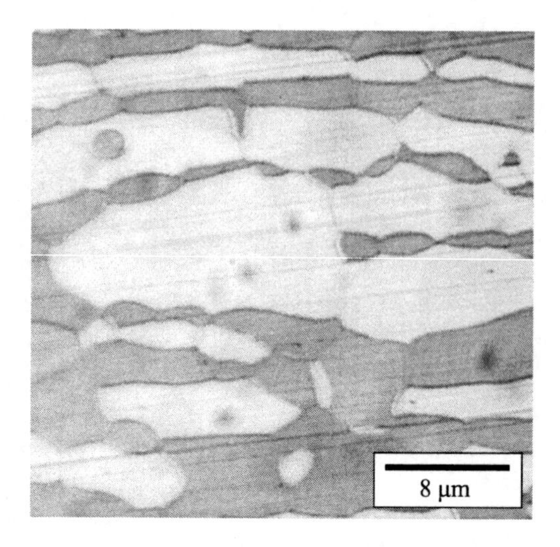

Figure 9.2 An example of a duplex stainless steel sample, microstructure of duplex stainless steel SAF 2205 containing about 0.003% carbon and 22.55% chromium (bright: austenite, dark: ferrite) [11]

Pigging has always been advised by experts [9, 12, 13] as a very feasible way to keep the system clean and to manage the possibility of MIC.

9.2.2 Use of Ultraviolet Radiation

Ultraviolet (UV) can be defined as a physical process in which the targeted organism(s) are not killed but their genetic material (DNA) is altered so that their production is prevented [14]. While some investigators have addressed UV application as an "alternative to biocides" [15], perhaps due to the efficiency of UV in a 99.9999% reduction in viable bacterial numbers [16], it has been reported that only on surfaces directly irradiated by UV may biofilm growth be prevented, so that as soon as non-living particles can shield the micro-organisms from UV by adhering onto the surface, the growth can be re-started [17]. Also, due to the poor penetrating power of ultraviolet light, this method has been reported to affect the planktonic but not the sessile bacteria in biofilms [16].

Some of the shortcomings of UV treatment can be briefly addressed as follows [14]:

- The UV lamp can be covered with micro-organisms, thus decreasing the UV radiation that could be available for de-activation. For example, if the bacteriisum a spore-former, it may require a 10-times higher dose of the UV light to be reduced 90% in count compared to its non-spore former equivalent strain.
- Some micro-organisms, especially certain types found in wastewater treatment, are not inactivated by UV radiation, and this could be a mater of scrutiny (especially in drinking water applications).

9.2.3 Ultrasonic Treatment

Another method that may be useful is applying power ultrasound (UT). The possible mechanisms by which UT can affect MIC are explained as follows [18]: through UT, an acoustic pressure is produced that induces cavitation bubbles in the liquid. When these bubbles later collapse, the high and intense pressures (in the order of hundreds of atmospheres) and temperatures (in the range of thousands of degrees) thus locally generated will have two detrimental effects: (1) they are capable of destroying the cells, and (2) by formation of chemical species such as hydrogen peroxide (which has biocidal effects) and hydroxyl radicals, the chemistry of the environment becomes very hostile to micro-organisms.

Ultrasonic energy has been reported as having "good efficacy" [17] and being a "promising method [against] soft biofilms" [19]. However, the feasibility of applying this method for mitigation of MIC depends on [18] the generation of enough cavitational forces to kill large enough numbers of MIC-assisting bacteria so that the regrowth[3] is low enough to ensure minimisation of corrosion.

Ultrasonic treatment may destroy the underlying material and be restricted to surfaces where UT can be applied [19].

9.3 Chemical Treatments

Using biocides is the most profound characteristic of chemical treatment. Biocide, literally meaning "killer of living [things]", can be divided into two large categories, oxidising and non-oxidising biocides. Oxidising biocides penetrate and destroy the bacterial cells, whereas non-oxidising biocides penetrate the biofilm and damage the cell membrane or destroy the mechanisms used by the micro-organism to process energy [20].

In the literature of chemical treatment of MIC by biocides, a very commonly used term for a biocide is "broad-spectrum". That means that the "broad-spectrum

[3] "Regrowth", "aftergrowth" or "recovery" all refer to rapid returning of biofilms back immediately after a biocidal treatment. There could be five reasons for regrowth: (1) if the remaining biofilm still has enough viable organisms to let the bacterial community jump from "lag phase" – where a critical size of bacterial population is needed to arrive at rapid growth (or, log phase where the increase in bacterial population is very rapid) – then, after a shock treatment, the bacterial number on such surfaces increases skyrocket in comparison with a previously clean surface, (2) the remaining biofilm offers a "rough" surface to the planktonic bacteria that can use it more efficiently than a clean surface, thus facilitating formation of more sessile bacteria, (3) biocides like chlorine may not be able to penetrate deep enough to affect the biofilm cells, in this case, while chlorine removes the outer cells and EPS, after chlorination stops, the "deep-down" cells will have a better access to nutrients so that their growth is enhanced, (4) the surviving "deep-down" cells will start to rapidly create EPS to counteract the effect of chlorine and (5) if there are micro-organisms that could be "less susceptible" to a biocidal treatment, they can rapidly proliferate between biocide treatment programs. See [19].

biocide" must be able to kill as many diverse types of micro-organisms and as many of the same micro-organism as possible. In other words, if a certain biocide is capable of killing both bacteria and fungi, it is a broader-spectrum biocide than a biocide that just kills bacteria. In the same way, if a biocide can kill several types of a certain bacterium, it is broader-spectrum biocide than the one that kills just one type of the same bacterium. Some of the biocide selection criteria are as follows [21]:

- The type of micro-organisms involved
- The prior operating history of the system
- The type of process cooling water system
- The chemicals being used for scale and corrosion control
- Chemical and physical characteristics of the water in the system
- Environmental limitations and restrictions

It is important to note that inhibitors are chemicals used mainly for the treatment of non-microbiological corrosion where biocides are used for killing micro-organisms. However, practices like adding inhibitors such as chromates in concentrations ranging from 50 to 1000 mg/L into systems where the pH of the system is kept in a non-scaling range by adding acid may render the corrosion inhibitor toxic to many of the micro-organisms capable of inducing MIC [22]. In this way, both non-microbial electrochemical corrosion and MIC can be treated.

9.3.1 Pros and Cons of Some Biocides

As mentioned earlier, biocides, by their effects, can be divided into two large groups: oxidising and non-oxidising biocides. Some examples of biocides mainly used in oil industry are presented below with their pros (+) and cons (–) [23–25]:

OXIDISING BIOCIDES

Chlorine

(+)

- Economical
- Broad-spectrum activity
- Effective
- Monitoring dosages and residuals is simple

(–)

- Hazard concerns for the operator
- Ineffective against biofilm bacteria

- Ineffective at high pH
- Inactivation by sunlight and aeration
- Corrosive to some metals
- Adverse effect on wood
- Feeding (dosing) equipment is costly and requires extensive maintenance
- Limitations imposed by environmental authorities on the discharge of chloramines and halomethanes

Chlorinating compounds (bleach [NaOCl], dry chlorine [$Ca(OCL)_2$])
(+)

- Circumvent the danger of handling chlorine
- As effective as chlorine

(–)

- Can cause scaling problems
- Expensive
- Larger quantities needed than when using gaseous chlorine

Chlorine dioxide (ClO_2) [26]
(+)

- pH-insensitive
- Good oxidising agent for biomass
- Tolerates high levels of organics
- Dissolves iron sulphides

(–)

- Special equipment is required for generation and dosing
- Toxic
- Expensive

Chloramines (like ammonium chloride)
(+)

- Good biofilm activity
- Good persistence in long distribution systems
- Has reduced corrosivity
- Low toxicity

(–)

- Ammonia injection is required
- Costs more than chlorine alone
- Poor biocidal properties compared to free chlorine [27]

Bromine

(+)

- More effective than chlorine at higher pH
- Broad spectrum activity on bacteria and algae over a wider pH range than hypochlorous acid
- Bromamines are environmentally less objectionable and less reactive with hydrocarbons, *etc.*, reducing the production of halomethane

(–)

- Similar to chlorine compounds
- Expensive

Ozone[4]

(+):

- A natural biocide, effective as a detachment agent against sessile bacteria on stainless steel surfaces
- Advantages similar to those of chlorine
- Non-polluting and harmless to aquatic organisms

(–)

- Like chlorine, it is affected by pH, temperature, organics, *etc.*
- Its oxidising effect does not resist throughout the system, so ozone is used in small systems or specific sites within larger systems
- Ozone must be generated on-site, requiring investment for installation and running the equipment

Sodium and Hydrogen Peroxides[5]

(+)

- Used as a sanitising agent
- Have many of advantages as ozone

(–)

- Require high concentrations and extensive contact time [to kill the micro-organisms]
- Cheaper and more safe than ozone
- Careful use not to stimulate corrosion

---

[4] One of the chemicals that in the role of a nutrient supports the growth of micro-organisms is assimilable organic carbon (AOC), which is a fraction of the organic matter that naturally exists in water. When ozone is added as a pert of an ozonation process, it increases AOC as a result of breaking up organic carbon large molecules into smaller molecules (see [28]). In other words, using ozone may kill the bacteria but, if not treated with intensive care, could cause regrowth promptly due to making organic matter more available to the micro-organisms.

[5] Biocidal effect of hydrogen peroxide may be due to it providing other alternative cathodic reduction in addition to oxygen reduction, thus enhancing the possibility of ennoblement, see [29].

NON-OXIDISING BIOCIDES

Aldehydes

1. Formaldehyde (HCHO)

(+)

- Economical

(–)

- Suspected of being a carcinogen
- High dosages are required
- Reacts with ammonia, hydrogen sulfide, and oxygen scavengers

2. Glutaraldehyde

(+)

- Broad-spectrum activity
- Relatively insensitive to sulphide
- Compatible with other chemicals
- Tolerates soluble salts and water hardness

(–)

- It is de-activated by ammonia, primary amines, and oxygen scavengers

3. Acrolein

(+)

- Broad-spectrum activity
- Penetrates deposits and dissolves sulfide constituents
- In highly contaminated waters, it is generally more economical/cost-effective than chlorine.
- No particular environmental hazards

(–)

- Difficult to handle
- Reactive with polymers, scavengers, and violently reacting with strong acid and alkalis
- Potentially flammable
- Highly toxic to humans

Amine-type compounds

1. Quaternary amine compounds

(+)

- Broad-spectrum activity
- Good surfactancy
- Persistence
- Low reactivity with other chemicals

(–)

- Inactivated in brines
- Foaming
- Slow acting

2. Amine and diamine
(+)

- Broad-spectrum activity
- Have some inhibition properties
- Effective in sulphide-bearing waters

(–)

- React with other chemicals, particularly anionics
- Less effective in waters with high levels of suspended solids

Halogenated compounds
1. Bronopol
(+)

- Broad-spectrum activity
- Low human toxicity
- Ability to degrade

(–)

- Available as a dry chemical
- Breaks down at high pH

2. DBNPA
(+)

- Broad-spectrum activity
- Fast-acting and effective (at a pH above 8.0, it must be used for quick kill situations)
- No apparent difficulties related to effluent discharge with these materials when applied as recommended

(–)

- Expensive
- Affected by sulphides
- Must be adequately dispersed to ensure effectiveness (due to low solubility in water)
- Although effective against bacteria at low concentrations, higher concentrations are required to control most algae and fungi, making them less cost-effective
- Overfeeding causes foaming and skin contact problems

Sulphur compounds

1. Isothiazolone[6]

(+)

- Broad-spectrum activity
- Compatible with brines
- Good control of many aerobic and anaerobic bacteria (such as anti-sessile bacteria) and have activity against many fungi and algae at acidic to slightly alkaline pHs
- Low dosages are required
- Degradable

(–)

- Cannot be used in sour systems
- Expensive
- Less cost-effective when the system contains significant amounts of sessile or adhering biomass. In such cases, the use of a penetrant/biodispersant enhances the effectiveness of the biocide
- Extreme care required because of potential adverse dermal effects; automated feeding systems are strongly recommended

2. Carbamates (alkyl thiocarbamates)

(+)

- Effective against SRB and spore formers
- Effective in alkaline pH
- Useful for polymer solutions

(–)

- High concentrations are required
- React with metal ions and other compounds

3. Metronidazole (2-methyl-5 nitroimidazole-1-ethanol)

(+)

- Effective against SRB
- Compatible with other chemicals

---

[6] The most frequently used types of isothiazolone are 3:1 ratio 5-chloro-2-methyl-4-isothiazoline-3-one (CMI), 2-methyl-4-isothiozolin-3-one (MIT) (see [30]), and also 4,5-dicholo-2-n-octyl-4-isothiazolin-3-one (DCOI) (see [31]). It has also been reported that [30] isothiazolones use a two-step mechanism to affect micro-organisms: step 1. takes minutes and it involves rapid inhibition of growth and metabolic activities, step 2, taking hours to become effective, is an irreversible cell damage that is basically a kill process and end up in loss of viability. An investigation (see [32]) reports that when isothiazolone molcule is degraded, it releases chlorine as chloride ion and "not as an organochlorine metabolite or by-product". Therefore, if chloride-induced corrosion is a concern in a system, it is prudent not to use this biocide or use it with high degree of care. In addition, it has also been reported that isothiazalones have an active-SH group, that in the presence of sulphide, it can be affected (see [33]).

(–)

- It is specific to anaerobic organisms

Quaternary phosphonium salts (quats)
(+)

- Broad spectrum of killing activity and good stability. They are generally most effective against algae and bacteria at neutral to alkaline pH
- Low toxicity
- Stable and unaffected by sulphides

(–)

- Not effective fungicides at any pH
- Their activity is mostly reduced by high chloride concentrations, high concentrations of oil and other organic foulants, and by accumulations of sludge in the system
- Excessive overfeed of some types of quats may contribute to foaming problems, especially in open recirculating systems with organic contaminations

Another way of grouping biocides is in accordance with their mechanisms of action. In this way, the biocides are divided into two subgroups as seen in Figure 9.3.

Figure 9.3 illustrates an alternative way of looking at how biocides can be effective through their "mechanisms of action". With regard to one of these biocidal chemicals, silver, an important note must be said here; sometimes silver is recommended as a biocide to industrial inquirers who are not allowed to use copper or mercury, obviously for environmental concerns. It has been reported that although

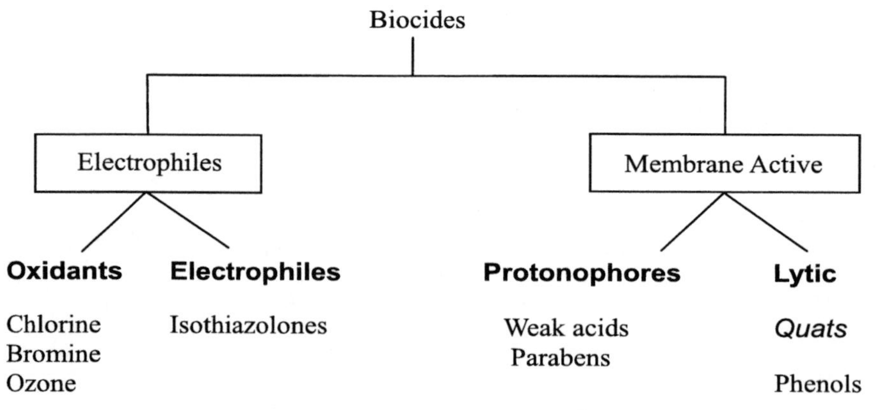

Figure 9.3 A brief of industrial biocides' mechanisms of action[7]

[7] See Footnote 8, [31].

Table 9.1 Comparison of antibiotic and biocide resistance modes

Characteristics of developing resistance to ***Antibiotics*** by bacteria	Characteristics of developing resistance to ***Biocides*** by bacteria
Antibiotics act by selective biochemical blocking of important binding sites of bacteria	Biocides act by mechanisms such as precipitation of proteins, solubilisation of lipids, *etc.* with attack on multiple sites in the bacterial cell simultaneously

silver is a "killer" to micro-organisms at very low concentrations (less than 10 μg/L), within a few weeks, the micro-organisms become not only tolerant but also start to multiply in the presence of concentrations as high as 1 mg/L of Ag ions [17].

Talking about developing resistance to some biocidal agents, we should explain a misunderstanding. It may be believed that bacteria change and modify their genetic features by undergoing periodic mutation, therefore after a period of time, they become resistant to a given biocide. Al-Hashem *et al.* [34] differentiate between adaptation and developing resistance to a biocide from an antibiotic. Theses differences can be summarised as in Table 9.1.

From the table it may be observed that bacteria would need to alter the structure of every protein in the cell to enable it to become resistant to a biocide. This would require that the bacteria would need a large number of mutations at the same time or in a short time to become resistant to a biocide. Such patterns of mutations, however, must occur over time spans much longer than what is normally available in an industrial system. The main reasons that explain why a biocide that seemed to be working previously is not working any more can be summarised as follows [34]:

1. change in biocide dosing regime
2. change of the manufacturer of the biocide
3. change of factors such as the system's temperature and pH
4. lack of biocide use optimisation so that it was effective on the small-sized initial bacteria population but with an increase in size and biological activity over time, the initial dosage of the biocide has proven to be ineffective.

9.3.2 A Note on Dual Biocide Treatment

It must be noted that physiological resistance of biofilms to oxidising biocides is much less pronounced than for non-oxidising biocides. Combinations of both oxidising and non-oxidising biocides in one treatment help to offset the physiological resistance of biofilms because of their dual mechanism of action [35].

Almost in the same way that in composite materials, the properties of the components (phases) are combined to give a better result, dual biocide treatment may also prove to be useful. For example, by combining the good killing features of an aldehyde with high penetration abilities of a quaternary amine, the poor penetration characteristics of the aldehyde and low killing efficiency of the amine are compromised [34].

However, Al-Hashem *et al.* [34] reported another example of dual biocide treatment where a high concentration batch treatment by a biocide was followed by a low concentration of continuos biocide treatment. The first step had been introduced to reduce the numbers of the bacteria (batch treatment) and keep it low (continuous treatment). Those authors, after not finding such treatments feasible enough both in terms of the extra time to be allocated for each treatment and the costs of the chemicals and facilities, prefer a batch dose of high concentration with special consideration of factors such as the frequency of biocide application and pre-treatment of the water entering into the system.

Another example [16] is using chlorinated (or brominated) compounds with surfactants to oppose biofilm formation. However, removal of the surfactants after the application can be a problem, for example in terms of the volumes of water needed to rinse them.

It seems that no matter how one interprets dual biocide treatment, either in terms of using two non-oxidising biocides or a combination of an oxidising and a non-oxidising biocide or even using different regimes and concentrations of the same biocide, these all depend on factors such as biocide selection, system requirements, and the economy of the application and post-treatment concerns. Therefore, although dual biocide treatment can be advisable, ignoring the factors just mentioned may result in a practice which will be hardly applicable.

9.4 Electrochemical Methods

It may appear a little strange to categorise items such as cathodic protection and coating under electrochemical methods. However, it will make sense when we think of these methods in terms of their effects on building up an electrochemical cell [see Chapter 1, electrochemical triangle]. In other words, coating is mainly replacing the role of electrolyte by separating electrodes (anode and cathode) from finding a medium through which electron and ions can be transferred. In the same way, by applying cathodic protection, the electrons lost from the metal during anodic reactions are provided by the CP system, and thus the role of anode becomes less important.

9.4.1 Cathodic Protection (CP)

The cathodic protection criterion of –0.95 V (Vs Cu-$CuSO_4$ reference electrode) to protect steel against SRB-induced MIC first appeared as the result of thermodynamic considerations in 1964 by Hovarth and Novak [36, 37] to be later experimentally verified by Fischer in the early 1980s [36, 38].

While the "–950 mV" criterion has been widely used, there are reports that show this criteria is not as straightforward as it may seem. Two such reports were cited in Chapter 4, where two examples of investigations done in the early and the late 1990s supported the idea that the "–950 mV" CP criterion may not actually be working the same everywhere. The results of an investigation of CP effects on pure iron surfaces in the presence of SRB [39] demonstrated that applying cathodic polarisation of –1070 mV Vs Cu-$CuSO_4$ has not been sufficient to prevent the growth of SRB.

The accepted theory to explain the feasibility of CP on MIC is that CP increases the local pH at metal/medium (water and/or soil) interface, thus causing the release of hydroxyl ions and decreasing the solubility of calcium and magnesium compounds [40]. This would result in the formation of calcareous deposits. It is this high pH generated by CP that has made some researchers speculate why CP is effective on MIC [41], as it is believed that micro-organisms cannot normally tolerate such high pH values. This is even though the presence (and not growth and vitality) of alkaliphilic micro-organisms in highly alkaline ($pH \geq 11$) media has been reported [42].

9.4.1.1 How Is CP Effective on MIC?

As pointed out earlier, a possible mechanism could be the chemical nature of the environment which is created after the application of CP in terms of increasing the local pH and inhibiting the bacterial reproduction of microbes [43] in such a high alkaline environment. But there are two seemingly rival theories in this respect, the electrostatic-chemical theory and the chemical bridge theory. We briefly explain these theories and interpretations below.

A. Electrostatic-Chemical Theory

In late 1990s it was reported by J.W. Arnold, a microbiologist then working at the ARS Poultry Processing and Meat Quality Research Unit at Athena, Georgia, U.S., that electropolished surfaces were much less vulnerable to biofilms build-up compared with the surfaces prepared by other methods such as polishing, sandblasting, and grinding [44]. A possible reason for observing such behaviour, it was theorised, could be the charge change induced by the electropolishing of the metal surfaces (that other polishing and surface treatments methods were not capable of), thus rendering the surface negatively charged. Therefore the bacteria, which can be taken as a charged particle due to their negative charge [45, 46], would not be able to attach themselves onto the surfaces easily.

If this interpretation is correct, then the negatively charged metallic surface (energised by CP and especially impressed current CP) repels the negatively charged bacteria as schematically shown in Figure 9.4.

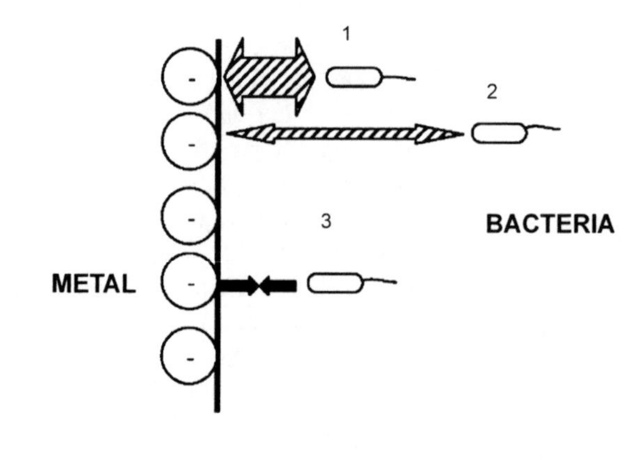

Figure 9.4 Possible interaction between a negatively charged surface and negatively charged bacteria. The power of the repulsion forces (shown as dashed) is schematically represented by the thickness of the arrows. For example, the magnitude of the repulsion forces against the bacterium ① is much larger than that of the bacterium ②. If the bacteria come close enough (0.4 nm or less [47]), then chances are that the interacting forces become attractive forces to let bacterial attachment onto the surface

In this model, the interaction between the negatively charged metallic surface and the negatively charged bacteria causes a lag phase before the chemical effect of CP starts to play a role. In other words, according to this model, which we call the electrostatic-chemical or EC model, when CP is on and the structure is energised, the repulsion forces thus produced would serve to keep the bacteria away from the structure as the negatively charged bacteria cannot be attracted to the structure unless the distance is in the order of nano-metres. While all this is happening, cathodic reactions are still on-going so that hydroxl ion release that occurs as a result of CP increases the pH locally, and the alkaline environment manages to affect the bacteria adversely. The outcome will be lowering the risk of corrosion. The EC mechanism is shown schematically in Figure 9.5.

As seen in Figure 9.4, when the bacteria come into contact with the metallic surface that, due to an induced current cathodic protection (ICCP), has already been negatively charged, the repulsion forces thus produced prevent the attachment of the bacteria onto the surface. This, in turn, would mean that the biofilm formation would be avoided. In Figure 9.5, possible stages involved in the CP that may be effective in reducing MIC are shown; the electrostatic effects are shown in Figure 9.5a and b. Taking the example of a pipe, it is schematically shown that due to CP, there is a fairly uniform charge distribution on the exterior wall of the pipe or a given segment of it (Figure 9.5a) where the pipe is surrounded by relatively non-uniformly distributed negatively charged bacteria (Figure 9.5b). The net effect will be repulsive forces that will push the bacteria away from the metallic surface. The chemical effect (Figure 9.5c, d) is that as the CP practice continues, the local concentration of protons (H^+) decreases by being used up in the cathodic reaction. By a further increase in pH, calcareous sediment formation is more assisted. As the local pH is too high, the micro-organisms that may be still adhering to the surface of the metal will die off.

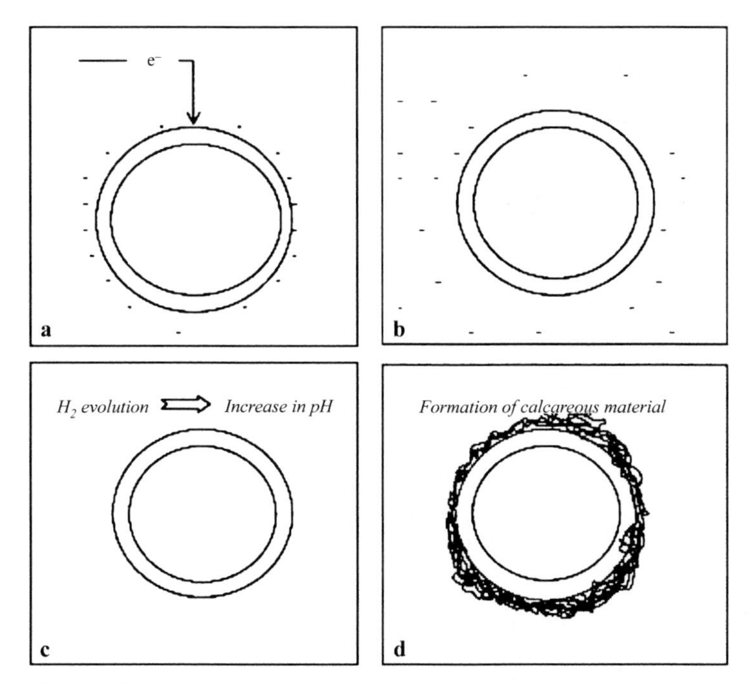

Figure 9.5 Possible EC effects on the ICCP of a pipeline [48]

Some examples of works favouring electrostatic repelling of negatively charged bacteria by the negatively charged metallic surfaces under CP have been quoted by Mains *et al.* [49].

An alternative theory that we call "chemical bridge theory" does not consider electrostatic forces of significant importance and rather relies on chemical binding, as will be discussed below.

B. Chemical Bridge Theory

Mains *et al.* [49], in trying to explain why applying CP to stainless and structural steel surfaces immersed in seawater can inhibit the settlement and attachment of aerobic bacteria to these surfaces, call the use of electrostatic repulsion theory in explaining such phenomena as being an "oversimplification". Instead, they propose an alternative mechanism. We call their proposed mechanism the chemical bridge theory, or CB.

Based on studies done on the adhesion of bacteria onto the surface of materials such as glass and tooth enamel and other studies addressed in their paper [49], their theory is shown schematically in Figure 9.6.

One question here is, why calcium and not anything else? One possible reason could be that polycations such as calcium ion or magnesium ion decrease electro-

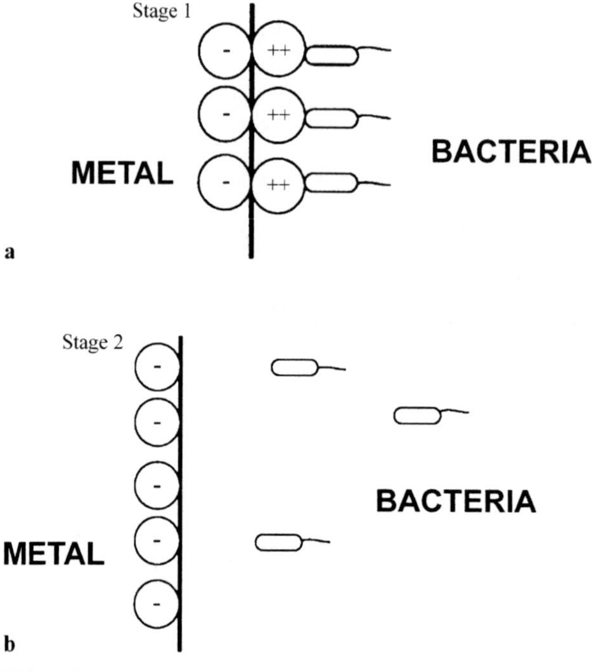

Figure 9.6 Stages involved in CB theory. **a** Stage 1: The bacteria use divalent ions such as calcium or magnesium to attach themselves onto the negatively charged metallic surfaces. **b** Stage 2: As CP increases, the local chemistry changes dramatically, resulting in a pH increase. This will turn the environment locally alkaline so that due to the precipitation of calcium and magnesium, these ions become unavailable to the bacteria

static repulsion during the primary stage of cell adhesion onto negatively charged surfaces [47]. However, the same study also noted that irreversible attachment of bacterial cells to solid surfaces could involve both monovalent and divalent cations [47].

It is not explained in the CB theory what will happen to the bacteria that have lost their "bridges". Therefore, we may assume that the fate of these bacteria will be left to the locally increasing alkalinity so that they may not be able to survive under those circumstances. In fact, another question that may come to mind is the possible events that can happen in between stages 1 and 2 from the standpoint of an increase in pH. Figure 9.7 shows the change in pH with regard to what is expected to happen according to CB theory:

The CB theory implies that there must be at least two increases in pH, one that is necessary to precipitate divalent ions such as calcium ions, which in Figure 9.7 has been marked by P_1 at the time T_1. Another increase in pH, marked by P_2 at time T_2, is when the pH becomes detrimental to the micro-organisms. At this time there is no evidence, to the best of the knowledge of the author, that indicates whether these pH increases are characteristically different from each other

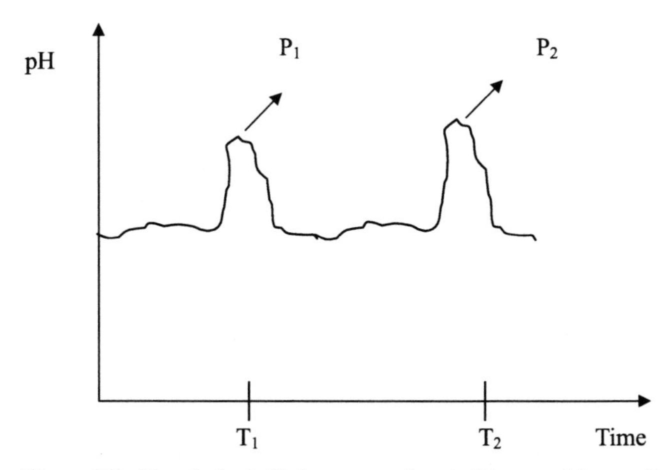

Figure 9.7 Hypothetical pH change over time (arbitrary scale) according to CB theory

($T_2 - T_1 > 0$) or the time difference is so infinitesimal ($T_2 - T_1 \approx 0$) that practically it is nil, suggesting that in Figure 9.7 there is only one peak, not two. This, in turn, means that one single pH rise is sufficient to both remove the bridging ions and make the environment hostile to the micro-organisms. However, it seems logical to imagine that if there is a lag time of $T_2 - T_1 > 0$, the micro-organisms repelled by the electrostatic forces would have a chance, however slim, to arrange themselves for a regrowth, should the lag time becomes long enough.[8] Basically, we can't even be sure about what the curve in Figure 9.7 may look like!

In fact, there are several other theories proposed to explain the mechanisms by which microbial adhesion on a surface can take place. These include, but may not be limited to, the Derjaguin Landau Verwey and Overback (DLVO) theory (which involves considering the effects of hydrophobicity and surface charge) and the theory of thermodynamics of attachment (which involves surface-free energy). These theories and their different aspects have been explained elsewhere [50] and we will not introduce those details here.

9.4.1.2 CP Criteria and Uncertainty in Design

The main purpose of this section is to show that the so-called –0.95 V criterion for having a secure CP against MIC may not be as straightforward in practice as it may seem in theory.

Around the world, for many engineering CP applications, only the effect of SRB and the –0.95 V criterion are considered when it comes to calculating the impact of

---

[8] Due to any possible reason ranging from poor practice of CP to irregularities in its application (which are not rare when it comes to the field conditions).

MIC on design. However, it is very important to realise that the –0.95 V criterion (and many more of its kind such as relating "certain" numbers of "certain" types of bacteria with a "certain" corrosion rate or assessment of MIC by the pit morphology) can, at their best, be regraded as the "minimum to expect". In other words, both the corrosion professionals and their clients must be educated to become aware of the existing shortcomings of these approaches and take them not as "solid rules" but "flexible guidelines". In the case of CP design and application, it may be a good idea to more accurately test the actual voltage at which, for a particular environment, CP can affect the microbial community by reducing their numbers or by keeping them "inactive" enough to not interfere with corrosion in any shape or mode. Needless to say, such practices can only occur in a perfect world where the cost of corrosion, in general, and MIC, in particular, is not an industrial joke.[9]

9.4.2 Coating

According to some surveys, "almost all" cases of corrosion of underground gas pipelines can be attributed to disbanded coatings [52]. Yet, an important issue with coating is that no matter how good a coating system is, as long as the application is not standard, the coating system may not be expected to perform well. In other words, although the idea of using protective coating is not new, there is still no ideal coating material that is adherent, coherent, completely non-porous, mechanically resistant to the hazards encountered during delivery, laying and backfilling, and chemically resistant to prolonged contact with all kinds of natural environments. Table 9.2 summarises the pros and cons of some coatings used for buried pipelines [53].

In addition to reportedly well-performing silicon-based coatings [54], new technologies that incorporate micro-fine copper flakes into an epoxy resin base to reduce biofilm adhesion [55] may seem promising for MIC-related corrosion issues. However it is still too soon to express an idea-positive or negative-on this subject.

Reportedly, some coats known as "soft-coat" or "semi-hard coatings" use vegetable oils. Needless to say how dangerous these coats could be with regard to MIC, as they would provide a "food" for the bacteria present in the untreated water coming into contact with them.

As some of the coatings could be polymer materials, it is useful to rank some frequently used polymers against microbial attack. Table 9.3 shows a selected series of such polymeric materials. It is advised, however, to study each related case individually and then make the decision as how one defines "stability" either with regard to their applicabilities such as mechanical properties or structural integrity for any particular case [56].

[9] See the last two paragraphs of the "introduction" of the paper by [51].

Table 9.2 Some features of commonly used coatings [53]

Coat Name	Advantages	Disadvantages
Coal tar-based	More stable and water proof than Asphaltic bitumen-based	Organic reinforcements to these coatings can be attacked and broken down by cellulose-decomposing microbes, carcinogenic, thus its use is banned in some countries
Asphaltic bitumen-based	Better to be reinforced with fibreglass	
Concrete	Alkalinity	Permeable to air, water, and stray currents, unless they are thick and hence expensive
Zinc coating on steel	With suitable thickness can prevent corrosion in neutral or alkaline soil for quite long time	Not to be used for acid conditions
Spray-applied zinc-aluminium coating	Promising	–
Lead coats	Good performance	Once the coating fails, rather rapid corrosion occurs
Plastic	Resistant to electrochemical corrosion	Bonding to metal
Fibreglass resin and epoxy resin coatings	Highly protective	Comparatively expensive

Table 9.3 Stability of some polymers to microbial attack [56][10]

Polymer	**Stability Ranking**
Polyethylene	*Very stable to medium stable*
Polypropylene	*Very stable to medium stable*
Polystyrene	*Very stable*
Polyurethanes	*Less stable*
Epoxy resins	*Very stable*

9.5 Biological Methods

In recent years, the feasibility of another method of MIC mitigation is being examined in which a certain type of bacteria is used against another type of bacteria. As

[10] Some of the micro-organisms that often attack plastics are *Pseudomonas aeruginosa*, as well as *micrococcus* and *bacillus* species. See [56].

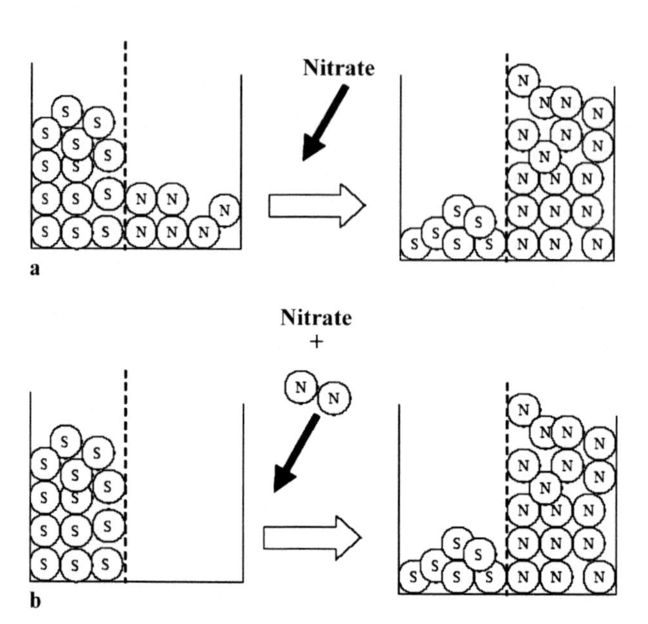

Figure 9.8 Explanation of two microbiological methods (bio-competitive exclusion and bio-augmentation) to mitigate MIC. **a** Bio-competitive exclusion by adding nitrate, nitrate-reducing bacteria (presented by circles with the letter N) will outnumber sulphate-reducing bacteria (shown by circles with the letter S). **b** Bio-augmentation Addition of *ex situ* grown nitrate-reducing bacteria and nitrate into a system that may have no "indigenous" nitrate-reducing bacteria

the reader may guess, the experiments are being done on the possibility of reducing MIC as induced by SRB.[11]

From some reports, it is known that some SRB (such as *Desulphovibrio desulphuricans* [58] and *Desulphovibrio gracillis* [59]) are capable of reducing nitrate.[12] Excluding such "weird" SRBs, some methods have been proposed and exercised to use nitrate-reducing bacteria (NRB) against SRB. Two examples of these methods are, bio-competitive exclusion and bio-augmentation [60]. The essential components of the definitions of these methods [60] are schematically presented as Figure 9.8.

Little *et al.* reported successful trials of bio-competitive exclusion as exercised on oil platforms where the corrosion rates werereduced by at least 50% [60]. On the other hand, with respect to bio-augmentation, although researchers such as Hubert *et al.* [61] and Bouchez *et al.* [62] have reported failures regarding the

[11] A possible, yet still theoretical, use of magnetic bacteria (Chapter 4) could be using them in a system contaminated with, say, SRB to corral the SRB and, literally speaking, "pushing" them to a spot under the effect of a magnetic field and then apply biocide to them. See [57].

[12] Nitrite has an inhibitory effect on SRB, mainly because: (a) nitrite is toxic to SRB, and with their nitrite reductase, the bacteria will produce a detoxifying reaction. The end result is that while the bacteria are still alive, no growth happens and their sulphate reduction activity will be inhibited. (b) Nitrite can directly affect the enzyme required for reducing sulphite to sulphide; see [63].

introduction of bacteria into natural mixed cultures, Zhu *et al.* [63] reported the simultaneous application of nitrate and denitrifying bacteria as "the most effective way" for controlling MIC induced by SRB. However, the research in this area is not yet complete.

9.6 Summary and Conclusions

Treatment of MIC can be done, with our present knowledge, in four categories: physical-mechanical, chemical, electrochemical, and biological. While all of these techniques have been refined and advanced with respect to just a few years ago, some of them, such as biological treatment of MIC, or suggestions for the use of coatings with nano-size copper flakes, are quite new. An important part of this section focused on cathodic protection and its effect(s) on MIC, pointing out that in the field of MIC there is hardly anything that has not been, or is not currently, under challenge. This, we hope, will once again highlight the vital need for more research and more communication among different disciplines of science and engineering and with the industrial world.

References

1. Javaherdashti, R., Corrosion Knowledge Management: How to deal with Corrosion as a Manager?, to be published.
2. Davies M, Scott PJB (1996) Remedial treatment of an occupied building affected by microbiologically influenced corrosion Mater Perform 35(6)54–57
3. Scott PJB, Davies M (1992) Microbiologically induced corrosion Civil Engineering 58–59
4. Guiamet PS, Gomez de saravia SG, Videla HA (1991) An innovative method for preventing biocorrosion through microbial adhesion inhibition. J Intl Biodeterioration & Biodegradation (43):31–35
5. McCoy WF (1998) Imitating natural microbial fouling control. Mater Perform 37(4):45–48
6. Source: http://en.wikipedia.org/wiki/Pigging#Images
7. Schmidt R (2004) Unpiggable pipelines – What a challenge for in-line inspection. Pigging Products and Services Association (PPSA) www.ppsa-online.com/papers.php
8. Verleun T (2004) Cleaning of oil & gas pipelines. Pigging Products and Services Association (PPSA) www.ppsa-online.com/papers.php
9. Jack TR (2002) Biological corrosion failures. ASM International
10. Archer ED, Brook R, Edyvean RG, Videla HA (2001) Selection of steels for use in SRB environments. Paper No. 01261. CORROSION 2001, NACE International
11. Javaherdashti R, Raman Singh RK, Panter C, Pereloma EV (2004) Stress corrosion cracking of duplex stainless steel in mixed marine cultures containing sulphate reducing bacteria. Proceedings of Corrosion and Prevention 2004 (CAP04), 21–24 November 2004, Perth, Australia
12. King RA (2007) Trends and developments in microbiologically induced corrosion in the oil and gas industry. MIC – An International Perspective Symposium. Extrin Corrosion Consultants, Curtin University, Perth, Australia, 14–15 February 2007
13. Scott PJB (2004) Expert consensus on MIC: Prevention and monitoring. Part 1. Mater Perform 43(3):50–54

14. Al-Majnouni AD, Jaffer AE (2003) Monitoring microbiological activity in a wastewater system using ultraviolet radiation as an alternative to chlorine gas. Paper No. 03067, CORROSION 2003, NACE International
15. Saiz-Jimenez C (2001) The biodeterioration of building materials. In: A practical manual on microbiologically influenced corrosion (Stoecket II JG, ed) 2nd edn, NACE International 2001
16. Mittelman MW (1990) Bacterial growth and biofouling control in purified water systems in biofouling and biodeterioration in industrial water systems. Proceedings of the International Workshop on Industrial Biofouling and Biocorrosion. Stuttgart, September 13–14 1990 (Flemming H-C, Geesey GG, eds) Springer-Verlag Berlin, Heidelberg 1991
17. Flemming H-C, Schaule G (1996) Measures against biofouling In: Microbially influenced corrosion of materials – scientific and engineering aspects (Heitz E, Flemming H-C, Sand W, eds) Springer-Verlag Berlin, Heidelberg
18. Pound BG, Gorfu Y, Schattner P, Mortelmans KE (2005) Ultrasonic mitigation of microbiologically influenced corrosion CORROSION 61(5)452–463
19. Flemming H-C (1990) Biofouling in water treatment in biofouling and biodeterioration in industrial water systems. Proceedings of the International Workshop on Industrial Biofouling and Biocorrosion. Stuttgart, September 13–14 1990 (Flemming H-C, Geesey GG, eds) Springer-Verlag Berlin, Heidelberg 1991
20. Grondin E, Lefebvre Y, Perreault N, Given K (1996) Strategies for the effective application of microbiological control to aluminum casting cooling systems. Presented at ET 96, Chicago, Illinois USA; 14–17 May 1996
21. Lutey RW (1995) Process cooling water. Section 3.3.6. In: Handbook of biocide and preservative use. Rossmore HW (ed) Blackie Academic & Professional (Chapman & Hall) Glasgow UK
22. Lutey RW (1995) Process cooling water. Section 3.2.4. In: Handbook of biocide and preservative use. Rossmore HW (ed) Blackie Academic & Professional (Chapman & Hall) Glasgow UK
23. Lutey RW (1995) Process cooling water. Section 3.4. In: Handbook of biocide and preservative use. Rossmore HW (ed) Blackie Academic & Professional (Chapman & Hall) Glasgow UK
24. Boivin J (1995) Oil industry biocides. Mater Perform 34(2):65–68
25. Videla HA,Viera MR, Guiamet PS, Staibano Alais JC (1995) Using ozone to control biofilms. Mater Perform (7):40–44
26. Cochran M, Extending ClO_2's Reach in Anti-microbial Applications, Special Chemicals Magazine, October 2004, www.speccheonline.com
27. Scott PJB (2000) Microbiologically influenced corrosion monitoring: Real world failures and how to avoid them. Mater Perform 39(1):54–59
28. Cantor AF, Bushman JB, Glodoski MS, Kiefer E, Bersch R, Wallenkamp H (2006) Copper Pipe Failure by Microbiologically Influenced Corrosion, *Materials Performance (MP)*, vol. 46, no. 6, pp. 38–41
29. Videla, H.A. (1995) Biofilms and Corrosion Interactions on Stainless Steel in Seawater, International Biodeterioration & Biodegradation, pp. 245–257
30. Williams TM (2006) The mechanism of Action of isothiazolone Biocides, Paper No. 06090, CORROSION 2006, NACE International, USA
31. Williams TM (2004) Isothiazolone Biocides in water Treatment Applications, Paper No. 04083, CORROSION 2004, NACE International, USA
32. Jacobson A, Williams TM (2000) The Environmental Fate of Isothiazolone Biocides, *Chimica Oggi,* Vol. 18, No. 10, pp. 105–108
33. King RA (2007) Microbiologically Induced Corrosion and biofilm Interactions, MIC – An International Perspective Symposium, Extrin Corrosion Consultants-Curtin University, Perth, Australia, 14–15 February 2007
34. Al-Hashem AH, Carew J, Al-Borno A (2004) Screening test for six dual biocide regimes against planktonic and sessile populations of bacteria. Paper No. 04748. CORROSION 2004, NACE International, USA

35. Ludensky ML, Himpler FJ, Sweeny PG (1998) Control of biofilms with cooling water biocides. Mater Perform 37(10):50–55
36. Kajiyama F, Okamura K (1999) Evaluating cathodic protection reliability on steel pipes in microbially active soils. CORROSION 55(1):74–80
37. Tiller AK (1986) Review of the European Research Effort on Microbial Corrosion between 1950 and 1984. In: Biologically induced corrosion. Dexter DC (ed) NACE–8, NACE Houston, Texas USA
38. Fischer KP (1981) Cathodic protection criteria for saline mud containing SRB at ambient and higher temperatures. Paper No. 110. CORROSION/81, NACE International, USA
39. de Romero MF, Parra J, Ruiz R, Ocando L, Bracho M, de Ricón OT, Romero G, Quintero A (2006) Cathodic polarisation effects on sessile SRB growth and iron protection. Paper No. 06526. CORROSION 2006, NACE International, USA
40. de Gonzalez CB, Videla HA (1998) Prevention and control. In: CYTED Ibero-American Programme of Science and Technology for Development, Practical Manual of Biocorrosin and Biofouling for the Industry, Subprogramme XV, Research Network XV.c. BIOCORR (Ferrari MD, de Mele MFL, Videla HA, eds) Poch&Industria Grafica SA, La Plata Bs.As., Argentina 1st edn
41. Geesey GG (1993) Biofilm formation. In: A practical manual on microbiologically-influenced corrosion Kobrin G (ed) NACE Houston Texas USA
42. Pedersen K (1999) Subterranean micro-organisms and radioactive waste disposal in Sweden. Engineering Geology (52):163–176
43. Stein AA (1993) MIC treatment and prevention. In: A practical manual on microbiologically-influenced corrosion Kobrin G (ed) NACE Houston Texas USA
44. Lee J (1998) Bacterial biofilms less likely on electropolished steel Agricultural Res p. 10
45. Percival SL, Knapp JS, Wales DS, Edyvean RGJ (2000) Metal and inorganic ion accumulation in biofilms exposed to flowing and stagnant water. Brit Corrosion J 36(2):105–110
46. Sreekumari KR, Nandakumar K, Kikuchi Y (2004) Effect of metal microstructure on bacterial attachment: A contributing factor for preferential mic attack of welds. Paper No. 04597. CORROSION 2004, NACE International
47. Geesey GG, Wigglesworth-Cooksey B, Cooksey KE (2000) Influence of calcium and other cations on surface adhesion of bacteria and diatomes: A review. Biofouling 15(1–3):195–205
48. Javaherdashti R (unpublished work) Mathematical justification of applying over-voltage in cathodic protection systems to avoid MIC
49. Mains AD, Evans LV, Edyvean RGJ (1991) Interactions between marine microbiological fouling and cathodic protection scale. In: Microbial Corrosion Proceedings of the 2nd EFC Workshop, Portugal 1991 (Sequeira CAC, Tillere AK, eds) European Federation of Corrosion Publications, Number 8, The Institute of Materials 1992
50. Habash M, Reid G (1999) Microbial biofilms: Their development and significance for medical devices-related infections. J Clin Pharmacol (39):887–898
51. Maxwell S, Devine C, Rooney F, Spark I (2004) Monitoring and Control of Bacterial Biofilms in Oilfield Water Handling Systems, Paper No.04752, CORROSION 2004, NACE International, USA
52. Li SY, Kim YG, Kho YT (2003) Corrosion behavior of carbon steel influenced by sulfate-reducing bacteria in soil environments. Paper No. 03549. CORROSION 2003, NACE International
53. Javaherdashti R, Vimpani P (2003) Corrosion of steel piles in soils containing SRB: a review. Proceedings of Corrosion Control and NDT, 23–26 November 2003, Melbourne, Australia
54. Wiebe D, Connor J, Dolderer G, Riha R, Dyas B (1997) Protection of concrete structures in immersion service from biological fouling with silicone-based coatings. Mater Perform 36(5):26–31
55. Metosh-Dickey CA, Portier RJ, Xie X (2004) A novel surface coating incorporating copper attachment. Mater Perform 43(10):30–34

56. Filip Z, Pommer E-H (contributors) (1992) Microbiologically influenced deterioration of materials. In: Microbiological degradation of materials and methods of protection. European Federation of Corrosion Publications, Number 9, The Institute of Materials
57. Javaherdashti R (1997) Magnetic bacteria against MIC, Paper No. 419, Corrosion 97, NACE International, USA
58. Dzierzewicz Z, Cwalina B, Chodurek E, Bulas L (1997) Differences in hydrogenese and APS-reductase activity between *Desulfovibrio desulfuricans* strains growing on sulphate or nitrate. Acta Biologica Cracoviensia Series Botanica (39:)9–15
59. Dunsmore BC, Whitfield TB, Lawson PA, Collins MD (2004) Corrosion by sulfate-reducing bacteria that utilize nitrate. Paper No. 04763. CORROSION 2004, NACE International, USA
60. Little B, Lee J, Ray R (2007) New development in mitigation of microbiologically influenced corrosion MIC – An International Perspective Symposium. Extrin Corrosion Consultants, Curtin University, Perth, Australia, 14–15 February 2007
61. Hubert C, Voordouw G, Arensdorf J, Jenneman GE (2006) Control of souring through a novel class of bacteria that oxidize sulfide as well as oil organics with nitrate. Paper No. 06669. CORROSION 2006, NACE International, USA
62. Bouchez T, Patureau D, Dabert, Juretschko S, Delgenes J, Molette R (2006) Ecological study of a bioaugmentation failure. As reported in [60]
63. Zhu XY, Modi H, Kilbane II JJ (2006) Efficacy and risks of nitrate application for the mitigation of SRB-induced corrosion Paper No. 06524. CORROSION 2006, NACE International, USA

Glossary

Aerobic bacteria: The bacteria that need oxygen to live.

Anaerobic bacteria: The bacteria that can live without oxygen.

Anode: The electrode at which oxidation occurs, from which the metal ions enter into the solution and the electrons flow away in the external circuit.

Biocide: A chemical which is lethal to any living thing, such as bacteria.

Broad-spectrum biocide: A biocide that can kill as many and as diverse micro-organisms type as possible.

Cathode: The electrode at which oxidation occurs, towards which the metal ions in the solution are attracted and the electrons in the external circuit flow.

Culture regime: In simple terms, the way a culture is refreshed. If the culture is not being refreshed at all, it is called a batch culture. Based on periodic or continuous refreshing, it is called semi- or continuous culture, respectively.

Culture: A chemical environment (culture medium) designed with certain organic and inorganic materials to support the growth of a certain type of bacteria (or other micro-organisms).

Electrolyte: The medium, normally a liquid, containing ions that in an electric filed migrate toward (away from) cathode (anode).

Facultative bacteria: The type of bacteria that can live either with or without oxygen.

IOB: The abbreviation for iron-oxidising bacteria.

IRB: The abbreviation for iron-reducing bacteria.

Mesophillic bacteria: The bacteria that grow best in room temperature.

Microbial corrosion: Also known as "microbiologically influenced corrosion", "microbiologically induced corrosion", "biocorrosion", or "MIC", "refers to an electrochemical corrosion type which is affected by micro-organisms such as certain bacteria. The effect could be accelerating the corrosion rate or decelerating it depending on many factors, including the dynamics of biofilms formation, the culture regime and the like.

Planktonic bacteria: The state in which the bacteria can freely float or swim in a body of water.

Sessile bacteria: The state in which the bacteria become motionless after being attached onto a surface.

SOB: The abbreviation for sulphur-oxidising bacteria.

SRB: The abbreviation for sulphate-reducing bacteria.

Thermophilic bacteria: The bacteria that grow best at temperatures above 50°C.

Index